Skeletal Muscle Biology and Therapeutics

Edited by Hilal Kalkan

Published in London, United Kingdom

Skeletal Muscle Biology and Therapeutics
http://dx.doi.org/10.5772/intechopen.1006770
Edited by Hilal Kalkan

Contributors
Abdelhay Addous, Ahmed Karim, Debprasad Dutta, Dongchao Lu, Hilal Kalkan, Jin Li, Krishnan Chakravarthy, Maja Green, Mohamed Marghich, Saliha Mharchi, Zena Alani

First published in London, United Kingdom, 2026 by IntechOpen
IntechOpen is the global imprint of INTECHOPEN LIMITED, registered in England and Wales, registration number: 11086078, 167-169 Great Portland Street, London, W1W 5PF, United Kingdom

For EU product safety concerns: IN TECH d.o.o., Prolaz Marije Krucifikse Kozulić 3, 51000 Rijeka, Croatia, info@intechopen.com or visit our website at intechopen.com.

British Library Cataloguing-in-Publication Data
A catalogue record for this book is available from the British Library

Skeletal Muscle Biology and Therapeutics
Edited by Hilal Kalkan
p. cm.

This title is part of the Physiology Book Series, Volume 35
Topic: Human Physiology
Series Editor: Tomasz Brzozowski
Topic Editor: Kunihiro Sakuma
Associate Topic Editors: Kotomi Sakai and Muneshige Shimizu

Print ISBN 978-1-83634-395-0
Online ISBN 978-1-83634-394-3
eBook (PDF) ISBN 978-1-83634-396-7
ISSN 2631-8261

If disposing of this product, please recycle the paper responsibly.

IntechOpen Book Series
Physiology
Volume 35

Aims and Scope of the Series

Modern physiology requires a comprehensive understanding of the integration of tissues and organs throughout the mammalian body, including the cooperation between structure and function at the cellular and molecular levels governed by gene and protein expression. While a daunting task, learning is facilitated by identifying common and effective signaling pathways mediated by a variety of factors employed by nature to preserve and sustain homeostatic life. As a leading example, the cellular interaction between intracellular concentration of Ca+2 increases, and changes in plasma membrane potential is integral for coordinating blood flow, governing the exocytosis of neurotransmitters, and modulating gene expression and cell effector secretory functions. Furthermore, in this manner, understanding the systemic interaction between the cardiovascular and nervous systems has become more important than ever as human populations' life prolongation, aging and mechanisms of cellular oxidative signaling are utilised for sustaining life. Altogether, physiological research enables our identification of distinct and precise points of transition from health to the development of multimorbidity throughout the inevitable aging disorders (e.g., diabetes, hypertension, chronic kidney disease, heart failure, peptic ulcer, inflammatory bowel disease, age-related macular degeneration, cancer). With consideration of all organ systems (e.g., brain, heart, lung, gut, skeletal and smooth muscle, liver, pancreas, kidney, eye) and the interactions thereof, this Physiology Series will address the goals of resolving (1) Aging physiology and chronic disease progression (2) Examination of key cellular pathways as they relate to calcium, oxidative stress, and electrical signaling, and (3) how changes in plasma membrane produced by lipid peroxidation products can affect aging physiology, covering new research in the area of cell, human, plant and animal physiology.

Meet the Series Editor

Prof. Dr. Thomas Brzozowski works as a professor of Human Physiology and is currently a Chairman at the Department of Physiology and is V-Dean of the Medical Faculty at Jagiellonian University Medical College, Cracow, Poland. His primary area of interest is physiology and pathophysiology of the gastrointestinal (GI) tract, with a major focus on the mechanism of GI mucosal defense, protection, and ulcer healing. He was a postdoctoral NIH fellow at the University of California and the Gastroenterology VA Medical Center, Irvine, Long Beach, CA, USA, and at the Gastroenterology Clinics Erlangen-Nuremberg and Munster in Germany. He has published 290 original articles in some of the most prestigious scientific journals and seven book chapters on the pathophysiology of the GI tract, gastroprotection, ulcer healing, drug therapy of peptic ulcers, hormonal regulation of the gut, and inflammatory bowel disease.

Meet the Volume Editor

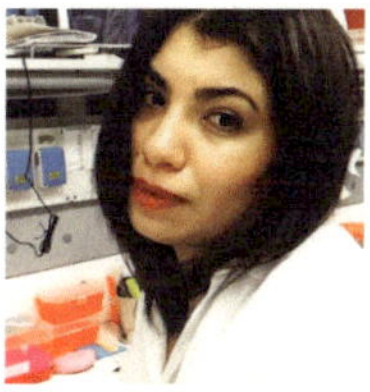

Hilal Kalkan is a postdoctoral researcher at Laval University in Quebec City, Canada, specializing in skeletal muscle biology, immunometabolism, cardiology, and the endocannabinoid system. Her research significantly enhances our understanding of the molecular mechanisms involved in muscle adaptation, inflammation, regeneration, and disease. She focuses particularly on communication between the gut and muscles, as well as on lipid signaling pathways. Dr. Kalkan completed her doctoral training in muscle biology and has published numerous peer-reviewed articles in prestigious journals, including a notable cover-featured piece in EMBO Molecular Medicine. Beyond her research efforts, she actively participates in scientific peer review, scholarly publishing, and interdisciplinary collaborations that connect fundamental muscle biology with translational and therapeutic innovations.

Contents

Preface

Skeletal muscle is a dynamic, multifaceted tissue essential for movement, posture, respiration, and metabolism, and functions as an active endocrine and immunometabolic organ. The study of skeletal muscle has evolved beyond classical physiology to encompass molecular signaling, regeneration, aging, and translational therapeutics.

The edited volume *Skeletal Muscle Biology and Therapeutics* aims to synthesize current perspectives on the biological principles governing skeletal muscle function and the therapeutic opportunities emerging from this field. The chapters explore the mechanisms of muscle development, repair, and disease, highlighting strategies to preserve or restore muscle health.

This multidisciplinary volume emphasizes the intricate regulatory networks that influence skeletal muscle, including metabolic pathways and inflammatory processes. Understanding these connections is crucial for clarifying disease mechanisms and identifying meaningful therapeutic targets.

The book addresses foundational aspects of muscle biology alongside contemporary topics, including regenerative medicine and the impact of lifestyle factors on muscle health. It aims to bridge mechanistic insights with therapeutic relevance, serving as a scientific reference for researchers, clinicians, and students across various biomedical disciplines.

I extend my gratitude to the contributing authors and the editorial teams who made this volume possible. We hope that *Skeletal Muscle Biology and Therapeutics* will provide a timely overview of the field's progress and inspire new research directions.

Hilal Kalkan
Department of Medicine,
Laval University,
Québec City, Canada

Section 1

Integrative Perspectives in Skeletal Muscle Biology

Chapter 1

Skeletal Muscle Biology and Therapeutics: From Molecular Mechanisms to Clinical Translation

Maja Green and Krishnan Chakravarthy

Abstract

Skeletal muscle is a metabolically active tissue essential for movement, energy regulation, and physiological balance. Muscle dysfunction resulting from aging, disuse, neuromuscular disease or critical illness is linked to significant morbidity, reduced mobility, and diminished quality of life. This chapter presents an integrated overview of skeletal muscle biology, focusing on molecular pathways that govern development, maintenance, and regeneration. Key signaling cascades, including PI3K/Akt/mTOR and FOXO, along with myogenic transcription factors such as MyoD and Myf5, are discussed in the context of fiber-type specification, hypertrophy, and atrophy. The chapter examines the roles of satellite cells, mitochondrial dynamics, inflammatory responses, and epigenetic regulators in muscular adaptation and repair. Therapeutic strategies range from pharmacologic agents and gene editing tools to stem cell–based interventions, neuromuscular electrical stimulation, and lifestyle modifications involving nutrition and exercise. Special attention is given to translational barriers and the disconnect between preclinical models and clinical practice. Emerging tools such as multi-omics profiling, artificial intelligence and biomarker discovery, are identified as promising for tailoring interventions and enhancing patient-specific treatment outcomes. Collectively, these innovations support the development of precision medicine approaches for muscle disorders. By bridging foundational biology with cutting-edge therapeutic options, this chapter offers a translational framework aimed at restoring skeletal muscle function across a range of clinical conditions and patient populations.

Keywords: skeletal muscle, myogenesis, neuromuscular stimulation, regenerative therapy, Skeletal muscle regeneration, muscle atrophy, myogenic signalling pathways, translational therapuetics

1. Introduction

Skeletal muscle constitutes approximately 40% of total body mass and is the largest organ system in the human body. It plays an indispensable role in voluntary movement, postural control, thermogenesis, glucose homeostasis, and systemic protein turnover. In addition to its mechanical function, skeletal muscle serves as a critical endocrine organ, secreting a wide range of myokines that influence metabolic, immunologic, and cardiovascular processes [1]. The preservation of skeletal muscle mass and contractile integrity is therefore essential for maintaining physiological resilience across the lifespan.

Muscle dysfunction and degeneration are central pathophysiological features in a broad spectrum of clinical conditions. Sarcopenia, cachexia, muscular dystrophies, and disuse atrophy represent distinct but often overlapping syndromes that contribute to frailty, impaired mobility, prolonged hospitalization, and increased mortality [2]. Moreover, skeletal muscle degradation is a significant predictor of poor outcomes in patients with cancer, sepsis, heart failure, chronic obstructive pulmonary disease, and critical illness [3]. Despite its clinical importance, effective strategies for preventing or reversing muscle wasting remain limited, underscoring an urgent need for mechanistically informed therapeutic approaches.

This chapter provides an integrative examination of skeletal muscle biology, spanning fundamental cellular and molecular mechanisms to translational and therapeutic innovations [4]. Key regulatory pathways involved in muscle development, hypertrophy, atrophy, and regeneration, such as the PI3K/Akt/mTOR axis, FOXO signaling, myogenic regulatory factors, and satellite cell activation, will be explored in detail. The chapter also highlights emerging therapeutic strategies, including pharmacologic agents, gene and cell-based therapies, electrical stimulation modalities, and precision-guided rehabilitation.

By bridging molecular science with clinical application, this chapter aims to elucidate the pathobiology of muscle disorders and provide a framework for developing and implementing targeted interventions. Particular emphasis is placed on the translational continuum, from in vitro and animal models to early-phase clinical trials and real-world applications. As the field of muscle therapeutics continues to evolve, integrating advances in biotechnology, regenerative medicine, and digital health will be essential for optimizing outcomes in both acute and chronic disease contexts.

2. Skeletal muscle structure and function

Skeletal muscle is a hierarchically organized, multinucleated tissue system that functions under the somatic nervous system to generate voluntary movement. Each muscle is composed of bundles of muscle fibers (myofibers), which are long, cylindrical syncytia formed by the fusion of myoblasts during embryogenesis. These myofibers are encased in endomysium, grouped into fascicles surrounded by perimysium, and collectively ensheathed by the epimysium, creating a mechanically and biochemically integrated contractile unit [5]. Embedded within this architecture is a dynamic population of satellite cells, fibroblasts, vascular cells, immune cells, and motor axons, each contributing to tissue homeostasis, plasticity, and regeneration.

2.1 Myofiber types and organization

Myofibers are classified based on their myosin heavy chain (MyHC) isoform expression, mitochondrial content, contractile velocity, and fatigue resistance. Type I (slow-twitch oxidative) fibers are rich in mitochondria, rely heavily on aerobic metabolism, and exhibit high fatigue resistance, making them suitable for postural and endurance activities. Type II fibers are further subdivided into IIa (fast oxidative-glycolytic), IIx, and IIb (fast glycolytic) based on their contractile properties and metabolic profiles. Type IIb fibers, for example, produce high force rapidly but fatigue quickly due to limited oxidative capacity. Fiber type composition varies across muscles and individuals and is modifiable in response to physiological demand, aging, and pathological stimuli, a process known as fiber type switching, orchestrated by calcium-dependent signaling, AMPK, PGC-1α, and mechanical loading cues.

2.2 Role of satellite cells and motor units

Satellite cells, the resident muscle stem cells located between the sarcolemma and basal lamina, are indispensable for postnatal muscle growth, repair, and adaptation. Under quiescent conditions, they express Pax7 but remain mitotically inactive. In response to injury or stress, satellite cells become activated, proliferate, and either differentiate into myogenic precursor cells (via MyoD and myogenin expression) or self-renew to replenish the satellite cell niche. Their coordinated activity is tightly regulated by Notch, Wnt, TGF-β, and FGF signaling, as well as niche interactions with the extracellular matrix and infiltrating immune cells.

Each myofiber is innervated by a single alpha motor neuron, and together they form a motor unit, the fundamental functional unit of muscle contraction. The size and firing pattern of motor units dictate the force and precision of muscle action. Neuromuscular junctions (NMJs) are specialized synapses that ensure efficient transmission of action potentials via acetylcholine release and postsynaptic receptor activation. Age, denervation, and disease can compromise NMJ integrity, leading to altered motor unit recruitment and muscle atrophy.

2.3 Metabolic and contractile properties

Skeletal muscle exhibits remarkable metabolic flexibility, switching between lipid oxidation, glucose uptake, and glycogenolysis based on energy demand and substrate availability. Mitochondria play a central role in ATP generation, reactive oxygen species (ROS) signaling, and apoptotic regulation. Myofibers also serve as a major reservoir for amino acids and glycogen, with systemic implications for metabolism during fasting, sepsis, or cachexia. Glucose uptake is mediated by insulin-dependent and independent pathways, including GLUT4 translocation, AMP-activated protein kinase (AMPK), and calcium-calmodulin signaling. Contractility arises from sarcomere shortening through actin-myosin cross-bridge cycling, regulated by calcium release from the sarcoplasmic reticulum (SR) and troponin-tropomyosin complex conformational changes [6]. Excitation-contraction coupling is initiated by depolarization of the sarcolemma and t-tubules, leading to voltage-gated calcium release via ryanodine receptors (RyR1). Abnormalities in calcium handling, mitochondrial function, or sarcomeric integrity can significantly impair force generation and contribute to contractile dysfunction in myopathies.

2.4 Interaction with neuromuscular and vascular systems

The function of skeletal muscle is inseparable from its integration with the peripheral nervous and vascular systems. Adequate perfusion ensures delivery of oxygen, glucose, and fatty acids, and facilitates thermoregulation and removal of metabolic byproducts such as lactate and carbon dioxide. Angiogenesis, driven by vascular endothelial growth factor (VEGF) and hypoxia-inducible factor-1α (HIF-1α), is tightly coupled with muscle hypertrophy and metabolic remodeling [7]. Endothelial cells also modulate satellite cell quiescence and activation through paracrine signaling. The neuromuscular system governs both voluntary contraction and reflexive responses through somatic afferent and efferent pathways. Motor neuron integrity is essential for sustaining muscle mass and function, and motor neuron loss in conditions such as amyotrophic lateral sclerosis (ALS) or spinal cord injury precipitates profound muscle wasting [8]. Moreover, chronic low-grade inflammation, oxidative stress, and neurovascular uncoupling can impair neuromuscular transmission and blood flow regulation, exacerbating disease progression in metabolic and degenerative conditions. Together, these interdependent structural and functional elements underscore the complexity of skeletal muscle as a system, not only as a contractile organ, but also as a hub of endocrine, metabolic, and regenerative activity. Understanding these interconnected pathways is essential for developing targeted interventions that restore muscle function in both localized and systemic disease states.

3. Molecular mechanisms of muscle development and maintenance

Skeletal muscle development and maintenance are governed by a tightly coordinated interplay of transcriptional regulators, signaling molecules, and metabolic pathways that determine fiber identity, growth, regeneration, and adaptability. Myogenesis begins during embryogenesis with the specification of mesodermal progenitors into the myogenic lineage, a process orchestrated by the myogenic regulatory factors (MRFs): Myf5, MyoD, myogenin, and MRF4 [5, 9]. Myf5 is expressed earliest and commits progenitor cells to the myogenic fate, while MyoD serves as a master regulator of terminal differentiation by activating a cascade of muscle-specific genes, including those encoding structural and contractile proteins such as myosin heavy chain (MyHC) isoforms. Myogenin is essential for myoblast fusion into multinucleated myotubes, and MRF4 contributes to postnatal muscle maturation and sarcomere assembly. The activity of these transcription factors is modulated by chromatin remodeling complexes, histone acetyltransferases, and myogenic co-activators such as MEF2, with additional fine-tuning provided by epigenetic regulators including histone deacetylases, DNA methyltransferases, and muscle-specific microRNAs (myomiRs) such as miR-1 and miR-206 [10]. Beyond transcriptional control, skeletal muscle homeostasis is strongly influenced by extracellular growth factors and intracellular signal transduction cascades. One of the most critical anabolic pathways involves insulin-like growth factor 1 (IGF-1), which activates the phosphoinositide 3-kinase (PI3K)/Akt signaling axis (**Figure 1**). Akt phosphorylates and inhibits FOXO transcription factors, thereby repressing the expression of the E3 ubiquitin ligases atrogin-1 (FBXO32) and MuRF1 (TRIM63), which are key mediators of proteasomal degradation and muscle atrophy. Downstream of Akt, the mammalian target of rapamycin (mTOR) complex 1 promotes protein synthesis by activating p70S6 kinase and inhibiting 4E-BP1, facilitating ribosomal biogenesis and translation

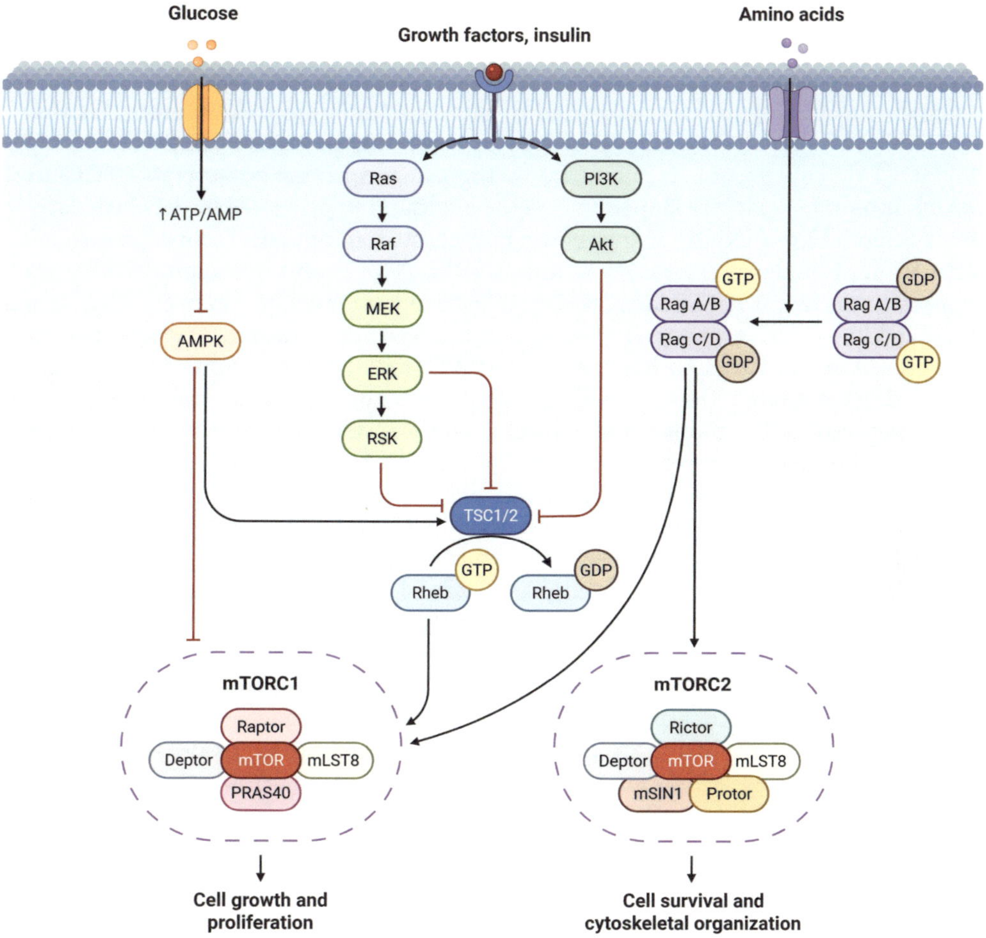

Figure 1.
mTOR signaling pathway in skeletal muscle regulation. This diagram illustrates the integration of nutrient availability (glucose and amino acids) and growth factor signals (e.g., insulin) through converging pathways that modulate mammalian target of rapamycin (mTOR) complex 1 (mTORC1) and complex 2 (mTORC2) activity. Growth factors activate pi3k/akt and ras/mapk signaling, both of which inhibit the TSC1/2 complex, leading to activation of rheb and mtorc1. Amino acids regulate mTORC1 through rag GTPases. Glucose and energy status influence mtorc1 indirectly via AMPK. mTORC1 promotes cell growth and proliferation through raptor, PRAS40, and mLST8. mTORC2, composed of rictor, Protor, and mSIN1, regulates cell survival and cytoskeletal organization. These pathways are essential for skeletal muscle hypertrophy, atrophy resistance, and metabolic adaptation.

initiation. This IGF-1/PI3K/Akt/mTOR axis is central to the hypertrophic response following resistance training and growth stimuli [11].

In contrast, growth differentiation factor 8, also known as myostatin, acts as a potent inhibitor of muscle growth through activation of the TGF-β/Smad2/3 signaling pathway. Myostatin binding to the ActRIIB receptor complex leads to Smad phosphorylation and nuclear translocation, where Smad2/3 suppresses myogenic gene expression and promotes protein catabolism. Inhibition of myostatin, either genetically or pharmacologically, has been shown to enhance muscle mass in both animal models and clinical trials [12]. Other regulatory inputs include fibroblast growth factors (particularly FGF2 and FGF6), which stimulate satellite cell proliferation via MAPK/ERK signaling, and Wnt proteins, which regulate myogenic lineage commitment and satellite cell self-renewal through canonical and non-canonical pathways [13]. Skeletal muscle is also capable of remarkable phenotypic plasticity in response to neuromuscular activity, loading, and metabolic stress. The calcineurin/NFAT pathway, activated by sustained calcium influx, promotes the expression of slow-twitch fiber genes by upregulating oxidative enzymes and type I MyHC [14]. This process is enhanced by the transcriptional co-activator PGC-1α, which integrates calcium-dependent signals with energetic cues to drive oxidative fiber remodeling. PGC-1α also governs mitochondrial biogenesis through co-activation of nuclear respiratory factors (NRF1, NRF2) and mitochondrial transcription factor A (TFAM), which collectively enhance the expression of genes involved in oxidative phosphorylation and mtDNA replication. AMPK, a cellular energy sensor activated by increased AMP/ATP ratio, complements this process by promoting catabolic efficiency, enhancing glucose uptake, and stimulating fatty acid oxidation, especially under conditions of metabolic stress or endurance exercise [15, 16]. Mitochondrial dynamics, including fission and fusion events mediated by proteins such as Drp1, Fis1, Mfn1, Mfn2, and OPA1, are essential for maintaining mitochondrial quality and adaptability. Dysfunctional mitochondria are removed through mitophagy, a process regulated by the PINK1/Parkin axis, ensuring the maintenance of bioenergetic capacity and redox homeostasis [17, 18]. Reactive oxygen species (ROS) generated during oxidative metabolism serve dual roles: at physiologic levels, they act as secondary messengers in redox-sensitive pathways such as MAPK, NF-κB, and FOXO, while excessive ROS accumulation leads to oxidative damage, mitochondrial dysfunction, and impaired regenerative signaling [19, 20]. These mechanisms are particularly relevant in aging, disuse, and chronic inflammatory states, where mitochondrial dysfunction contributes to reduced ATP production, elevated proteolysis, and impaired satellite cell function.

The cumulative action of these transcriptional networks, signaling pathways, and metabolic regulators establishes the dynamic equilibrium required for muscle maintenance across varying physiologic states. Disruption of these pathways, through genetic mutation, systemic disease, injury, or disuse, can precipitate muscle loss and functional decline [9]. A deeper understanding of these interconnected molecular mechanisms has not only elucidated the pathogenesis of muscle-wasting conditions but also enabled the rational design of targeted therapies aimed at restoring muscle mass, function, and regenerative capacity.

4. Muscle atrophy and regeneration

Skeletal muscle mass is maintained through a delicate balance between protein synthesis and degradation. When this balance shifts toward catabolism, whether due to mechanical unloading, neural loss, systemic disease, or aging, atrophy ensues [2]. Muscle atrophy is not merely a reduction in size, but a multifaceted biological response involving transcriptional reprogramming, proteolytic activation, mitochondrial dysregulation, and impaired regenerative signaling. The molecular signature of atrophy is characterized by a coordinated upregulation of catabolic pathways and downregulation of anabolic stimuli, often in response to altered endocrine, inflammatory, or neuromuscular inputs [21].

At the center of proteolytic degradation during atrophy is the ubiquitin–proteasome system (UPS), wherein muscle-specific E3 ubiquitin ligases, including atrogin-1 (FBXO32) and muscle RING finger 1 (MuRF1/TRIM63), catalyze the polyubiquitination of key myofibrillar proteins, targeting them for degradation by the 26S proteasome. The expression of these ligases is directly controlled by FOXO transcription factors, which translocate to the nucleus in the absence of inhibitory phosphorylation by Akt [22]. Under conditions of reduced IGF-1/Akt signaling, common in disuse, denervation, and systemic illness, FOXO1 and FOXO3a are de-repressed, driving the transcription of atrophy-related genes ("atrogenes") and tipping the balance toward net protein loss [23, 24]. In parallel, muscle atrophy engages the autophagy–lysosome system, particularly during prolonged stress or starvation. Autophagosome formation is upregulated via Beclin-1, LC3B, and BNIP3, allowing the selective degradation of damaged mitochondria, aggregated proteins, and sarcoplasmic components [25]. While autophagy is essential for cellular homeostasis, excessive or dysregulated activation can exacerbate muscle wasting. In conditions such as cancer cachexia, sepsis, and advanced sarcopenia, both UPS and autophagy pathways are upregulated simultaneously, leading to rapid and often irreversible tissue depletion [26]. The underlying causes of muscle atrophy are multifactorial. Disuse atrophy, seen in immobilization or microgravity, results from mechanical unloading and reduced neuromuscular activation. Denervation atrophy, resulting from peripheral nerve injury or motor neuron loss, disrupts trophic signaling and leads to fiber-type–specific degeneration, particularly in fast-twitch fibers. Cachexia, a hallmark of chronic systemic diseases such as cancer, heart failure, and COPD, is mediated by inflammatory cytokines including TNF-α, IL-1β, and IL-6, which activate NF- κ B and STAT3 pathways, enhance catabolism, and blunt anabolic signaling [27, 28]. Sarcopenia, the age-associated loss of muscle mass and function, is driven by hormonal decline (e.g., testosterone, GH, IGF-1), chronic low-grade inflammation, mitochondrial dysfunction, and impaired satellite cell responsiveness [29, 30].

Despite the diverse etiologies of muscle atrophy, regeneration remains possible through the activation of endogenous repair mechanisms. Central to this process is the activation of satellite cells, which re-enter the cell cycle in response to mechanical injury, inflammatory cues, and nitric oxide–dependent signaling. Upon activation, satellite cells express Pax7, proliferate under the influence of FGF2, and differentiate via MyoD and myogenin into myoblasts that fuse to existing myofibers or form de novo fibers. Successful regeneration depends on a finely tuned inflammatory response: an initial pro-inflammatory phase dominated by M1 macrophages facilitates debris clearance, while a subsequent M2-polarized anti-inflammatory phase supports myogenesis through the secretion of IL-10, TGF-β, and IGF-1 [31, 32]. Epigenetic mechanisms also play a pivotal role in regulating muscle regeneration. Histone modifications and DNA methylation modulate the accessibility of key myogenic promoters, while microRNAs provide post-transcriptional control over both satellite cell differentiation and fibrosis [33, 34]. For example, miR-1 and miR-206 promote myogenic differentiation by targeting Pax7 and HDAC4, whereas miR-21 and miR-29 modulate fibrotic remodeling by regulating collagen and TGF-β signaling components [35–37]. Dysregulation of these miRNAs in aged or diseased muscle contributes to defective regeneration and the accumulation of fibrotic matrix.

Skeletal muscle atrophy arises from complex and converging molecular programs involving ubiquitin-mediated proteolysis, autophagy, inflammatory signaling, and satellite cell dysfunction. The ability to regenerate following atrophic insult depends not only on the preservation of the satellite cell pool but also on the integration of appropriate environmental cues, epigenetic regulation, and the resolution of inflammation [38, 39]. These insights have significant implications for the design of targeted interventions aimed at mitigating muscle loss and restoring contractile function across a range of clinical contexts.

5. Therapeutic strategies

5.1 Pharmacologic approaches

Pharmacologic therapies for skeletal muscle dysfunction aim to restore the balance between protein synthesis and degradation, enhance regenerative potential, and attenuate the inflammatory and catabolic processes that drive muscle wasting. Several classes of agents have been developed or repurposed to achieve these objectives, including anabolic hormones, myostatin pathway inhibitors, and anti-inflammatory compounds. These interventions have shown varying degrees of efficacy across different etiologies of muscle loss, such as sarcopenia, cachexia, disuse atrophy, and genetic myopathies [40–42]. Anabolic agents remain the cornerstone of pharmacologic strategies designed to enhance muscle mass and strength. Testosterone and testosterone analogues, through androgen receptor activation, stimulate myofibrillar protein synthesis, increase satellite cell proliferation, and improve neuromuscular connectivity [43]. Their use is associated with measurable gains in lean body mass, particularly in hypogonadal men and older adults with sarcopenia. However, long-term administration raises concerns regarding cardiovascular risk, erythrocytosis, and prostate-related adverse effects, necessitating close clinical monitoring [44, 45]. Selective androgen receptor modulators (SARMs), such as enobosarm, aim to preserve the anabolic benefits of testosterone while minimizing androgenic side effects by targeting muscle and bone tissues selectively [46, 47]. Although initial trials have demonstrated improvements in lean mass and functional performance, questions regarding long-term safety and regulatory approval remain unresolved. Targeting the myostatin signaling axis represents another promising pharmacologic avenue. Myostatin (GDF-8), a member of the TGF-β superfamily, is a potent endogenous inhibitor of skeletal muscle growth. Inhibition of myostatin or its receptor complex (ActRIIB) leads to increased muscle fiber size and satellite cell activity. Multiple agents are under investigation, including neutralizing antibodies (e.g., stamulumab), soluble decoy receptors, and ligand traps [48, 49]. While these strategies have produced robust hypertrophic effects in preclinical models, translation to human disease has been challenging. In clinical trials involving patients with muscular dystrophies and cancer cachexia, myostatin inhibitors have demonstrated inconsistent effects on functional outcomes, and several programs have been discontinued due to lack of efficacy or safety concerns [50]. Ongoing efforts aim to refine patient selection and optimize dosing regimens to better capture their therapeutic potential. Chronic low-grade inflammation is a key driver of muscle catabolism in aging and chronic disease. As such, anti-inflammatory therapies have been investigated as adjuncts in preserving muscle mass and function. Non-steroidal anti-inflammatory drugs (NSAIDs) can blunt prostaglandin-mediated inflammation and reduce proteolytic signaling, but their use is limited by gastrointestinal and renal side effects, particularly in older populations. More targeted interventions include IL-6 antagonists, TNF-α inhibitors, and JAK/STAT pathway modulators, which aim to suppress specific inflammatory cytokines implicated in cachexia and sarcopenia [51, 52]. Although these therapies have shown promise in systemic inflammatory disorders, their application in muscle-wasting conditions is still under active investigation. Agents such as tocilizumab and etanercept have been evaluated in small-scale studies, but robust, phase III-level evidence remains scarce [53]. Additional pharmacologic candidates include β2-adrenergic agonists, such as clenbuterol, which stimulate hypertrophy through increased cAMP and protein synthesis, though cardiovascular risks and tolerance issues limit long-term use. Metabolic modulators, including AMPK activators, mTOR agonists, and nutraceuticals like leucine and HMB (β-hydroxy β-methylbutyrate), are also being explored for their anabolic or anti-catabolic effects. Collectively, pharmacologic strategies provide a mechanistic foundation for muscle preservation and enhancement but are often limited by off-target effects, heterogeneous responses, and the complexity of underlying pathophysiology [54]. As such, their utility may be maximized when combined with physical, nutritional, and regenerative interventions in a multimodal treatment paradigm tailored to individual patient phenotypes.

5.2 Gene and stem cell therapies

Gene and stem cell therapies represent rapidly evolving strategies aimed at correcting primary molecular defects, restoring regenerative potential, or modifying the muscle microenvironment in both inherited and acquired myopathies. The emergence of precision gene-editing tools, combined with advances in vector engineering and pluripotent stem cell technologies, has opened new avenues for addressing muscle pathology at its source [55]. CRISPR/Cas9 genome editing offers a highly specific and programmable system to modify disease-causing mutations in muscle cells. In monogenic disorders such as Duchenne muscular dystrophy (DMD), CRISPR has been used to excise mutated exons, restore the dystrophin reading frame, and partially reconstitute protein expression in preclinical models [56, 57]. Both in vivo and ex vivo strategies are under active development, with in vivo approaches leveraging adeno-associated virus (AAV) vectors to deliver Cas9 and guide RNAs directly to muscle tissue. AAV serotypes such as AAV8 and AAV9 exhibit tropism for skeletal and cardiac muscle and have shown promising transduction efficiency. However, challenges remain, including immune responses to viral capsids and Cas9, off-target mutagenesis, and the limited cargo capacity of AAV systems, which constrain the co-delivery of large transgenes or multiple regulatory elements [58]. In parallel, induced pluripotent stem cells (iPSCs) have provided an expandable and patient-specific platform for generating myogenic progenitors. iPSCs derived from skin or blood cells can be differentiated into satellite-like cells capable of engraftment and regeneration when transplanted into injured or dystrophic muscle. These cells retain the potential for gene correction prior to transplantation, enabling autologous cell therapy with reduced immunogenicity [59, 60]. Efforts to improve in vivo survival, migratory capacity, and myogenic fidelity are ongoing, with scaffold-based delivery systems and niche-modulating factors under investigation [61]. Despite substantial progress in preclinical studies, the clinical translation of gene and stem cell therapies for muscle diseases remains limited by delivery barriers, long-term safety concerns, and the need for scalable, GMP-compliant manufacturing. As these platforms evolve, integration with supportive technologies, such as AI-guided rehabilitation systems like EcoAI, may offer a synergistic approach by optimizing post-treatment muscle function, monitoring recovery trajectories, and tailoring neuromodulatory inputs to the regenerative timeline [62, 63].

5.3 Electrical and biophysical modalities

Biophysical interventions such as neuromuscular electrical stimulation (NMES) and functional electrical stimulation (FES) have long been utilized to preserve or restore muscle mass and function in settings of disuse, denervation, and motor impairment. These modalities deliver electrical impulses to peripheral motor nerves or directly to muscle fibers, inducing involuntary contractions that mimic voluntary movement, enhance local perfusion, and prevent atrophy [64]. NMES is commonly applied in intensive care settings, post-operative rehabilitation, and neuromuscular reeducation protocols. It improves mitochondrial oxidative capacity, promotes angiogenesis, and induces hypertrophic signaling via Akt/mTOR and calcium/calmodulin-dependent pathways [65]. FES, which is typically synchronized with task-specific movements, is especially beneficial in spinal cord injury, stroke, and other central neurologic disorders, where voluntary activation is impaired. It can support functional activities such as standing, stepping, or grasping through pre-programmed stimulation patterns coordinated with residual motor capacity [66]. More recently, biophysical technologies have expanded to include pulsed ultrasound, low-level laser therapy, and vibration platforms, each designed to modulate muscle tissue through non-electrical means. Pulsed ultrasound, for example, may enhance satellite cell proliferation, improve local circulation, and reduce fibrosis, though its mechanistic underpinnings remain incompletely defined [67]. The evolution of these modalities has culminated in the development of intelligent, closed-loop platforms such as the EcoAI system, which integrates NMES and TENS (transcutaneous electrical nerve stimulation) with real-time patient feedback and machine learning algorithms. Unlike traditional stimulation devices that operate with fixed, clinician-determined parameters, EcoAI continuously adapts waveform intensity, frequency, and pulse width based on patient-reported outcomes and physiologic data. By leveraging remote monitoring and cloud-based analytics, the system not only enhances patient adherence but also enables longitudinal tracking of muscle performance, neuromodulatory responsiveness, and therapeutic outcomes [62]. This represents a paradigm shift in electrical stimulation therapy, transitioning from static rehabilitation tools to intelligent, autonomous neuromuscular systems capable of delivering tailored interventions with clinical oversight. These electrical and biophysical modalities represent essential tools in muscle rehabilitation and are most effective when integrated into comprehensive, multimodal treatment strategies. Emerging digital platforms such as EcoAI enhance this paradigm by enabling precision-guided, adaptive neuromodulation that links molecular therapy to functional restoration.

5.4 Nutritional and exercise interventions

Nutritional optimization and targeted exercise remain foundational elements in the preservation and restoration of skeletal muscle mass and function. These interventions exert their effects through direct stimulation of muscle protein synthesis (MPS), enhancement of mitochondrial and capillary density, modulation of endocrine signaling, and reduction of systemic inflammation [68]. Their efficacy is influenced not only by the biochemical quality of inputs, such as amino acid composition and training intensity, but also by the temporal alignment of nutrient delivery, the consistency of physical activity, and the degree of patient engagement. Emerging technologies, including wearable sensors and remote patient monitoring (RPM) platforms, are now being leveraged to improve adherence, personalize interventions, and objectively track response to therapy in real-world settings. Adequate protein intake is essential for maintaining muscle anabolism, particularly in aging populations where anabolic resistance attenuates the MPS response to feeding. The recommended protein intake for older adults is now widely accepted to exceed 1.2 g/kg/day, with higher thresholds indicated in sarcopenic or catabolic states. Leucine, a branched-chain amino acid, plays a pivotal role in initiating MPS via mTORC1 activation and translation initiation. Supplementation with leucine-rich proteins or leucine analogs such as β-hydroxy β-methylbutyrate (HMB) has been shown to enhance muscle mass and reduce proteolysis, particularly in hospitalized and immobilized patients [69]. The timing of protein ingestion, ideally within a post-exercise anabolic window, further augments its effectiveness by exploiting exercise-induced sensitivity to amino acids. Resistance training remains the most potent non-pharmacologic stimulus for muscle hypertrophy. Mechanical loading activates intracellular signaling cascades, including PI3K/Akt/mTOR and focal adhesion kinase pathways, which drive protein synthesis and satellite cell activation. High-intensity, progressive overload regimens are particularly effective in stimulating type II fiber hypertrophy, improving neuromuscular coordination, and reversing disuse atrophy. In older adults and those with limited mobility, low-load, high-repetition protocols and blood flow restriction training offer viable alternatives. Endurance exercise, while less effective for hypertrophy, improves mitochondrial function, vascular density, and oxidative capacity [70, 71]. A key challenge in translating nutritional and exercise prescriptions into long-term outcomes lies in sustaining patient motivation and adherence. To address this, remote patient monitoring (RPM) technologies, particularly wearable activity trackers, smart scales, and app-integrated nutrition logs, are increasingly used to provide real-time feedback, reinforce behavioral change, and identify early signs of noncompliance or clinical deterioration. When integrated into clinician-facing dashboards or AI-driven therapeutic platforms such as EcoAI, RPM data can inform individualized adjustments to physical activity targets, dietary intake, and neuromodulatory interventions. These systems allow for dynamic risk stratification, outcome prediction, and proactive clinical engagement, thereby closing the loop between behavior, physiology, and intervention. The convergence of nutritional science, exercise physiology, and digital health presents a scalable framework for muscle rehabilitation [72, 73]. Tailoring interventions to an individual's metabolic profile, recovery trajectory, and motivational drivers, while continuously tracking outcomes through RPM, offers a path toward precision-guided, sustainable therapeutic delivery in both preventive and rehabilitative contexts.

6. Translational models and clinical trials

The path from molecular insight to therapeutic application in skeletal muscle disorders requires robust translational models and rigorously designed clinical trials.

Preclinical studies using animal models remain foundational for understanding disease mechanisms and evaluating candidate therapies. However, translational fidelity is limited by interspecies differences in physiology, regenerative capacity, and disease progression. As the field advances, the integration of biomarker-based endpoints, real-time monitoring, and artificial intelligence is redefining how efficacy and safety are assessed in both experimental and clinical settings. Animal models, particularly in rodents, have been central to the study of muscle development, atrophy, and regeneration. Models such as the mdx mouse for Duchenne muscular dystrophy, hindlimb suspension for disuse atrophy, and cardiotoxin-induced injury for regeneration research have been widely used to identify critical signaling nodes, including the IGF-1/PI3K/Akt axis, myostatin/TGF-β signaling, and the ubiquitin-proteasome system. However, the extrapolation of therapeutic effects from animals to humans remains challenging. Rodents possess a highly efficient regenerative response, minimal fibrosis after injury, and a high proportion of type II muscle fibers, characteristics that diverge significantly from human muscle physiology [74, 75]. Additionally, most rodent models lack the comorbidities, such as chronic inflammation, metabolic dysfunction, or multimorbidity, that frequently accompany muscle wasting in humans, limiting their translational relevance. Larger animal models, including dogs, pigs, and non-human primates, offer improved anatomical and physiological similarity but are constrained by cost, ethical considerations, and limited availability. Models such as the golden retriever muscular dystrophy (GRMD) dog more closely mirror the progressive weakness and fibrosis seen in human DMD, making them valuable for evaluating gene and cell therapies [76]. Nonetheless, even these models cannot fully replicate the clinical heterogeneity of human muscle disorders.

The incorporation of biomarkers has become increasingly important for improving translational efficiency and regulatory acceptance. Biomarkers for skeletal muscle disorders fall into several categories: molecular, imaging-based, physiological, and functional. Circulating markers such as creatine kinase (CK), myostatin, osteopontin, IL-6, and miR-206 provide dynamic readouts of muscle damage, inflammation, and regeneration. Imaging biomarkers, particularly quantitative MRI and ultrasound elastography, allow for the non-invasive assessment of muscle volume, fat infiltration, edema, and fibrosis [77, 78]. Electrophysiological parameters, such as compound muscle action potentials and electromyography, offer insight into neuromuscular transmission and recruitment. Functional biomarkers, including gait speed, handgrip strength, sit-to-stand tests, and patient-reported outcome measures (PROMs), are commonly employed in clinical trials to assess therapeutic benefit, though they are often influenced by motivation, comorbidity, and learning effects. The past decade has seen a surge in early-phase and pivotal clinical trials for muscle-targeted therapies, reflecting increased interest from academic, industry, and regulatory stakeholders [79]. Gene therapy trials using AAV vectors in DMD, including micro-dystrophin replacement strategies, have entered phase I/II and shown partial restoration of dystrophin protein in muscle tissue [80]. Trials of myostatin inhibitors (e.g., bimagrumab, domagrozumab) in sarcopenia and cachexia have yielded mixed results, with improvements in lean body mass but inconsistent functional outcomes [81]. Small-molecule agents such as SARMs, mTOR modulators, and nutritional supplements like HMB have also been evaluated with variable success [82]. Notably, the heterogeneity of inclusion criteria, reliance on subjective outcome measures, and relatively short intervention periods have limited the interpretability of many trials.

Recent studies have begun to incorporate RPM and AI-driven platforms such as EcoAI to enhance trial design and data quality. These systems enable continuous, real-world tracking of muscle function, adherence, and patient-reported outcomes, providing granular data that may complement traditional endpoints and reduce reliance on episodic in-clinic assessments. Such integration not only supports adaptive trial designs but also facilitates longitudinal analysis, risk stratification, and more responsive intervention strategies. As regulatory frameworks evolve to accommodate advanced biologics, digital therapeutics, and decentralized clinical trials, the integration of high-fidelity translational models, validated biomarkers, and technology-enabled endpoints will be essential for the efficient and ethical development of muscle-targeted therapies.

7. Challenges and future directions

Despite significant advances in the molecular understanding of skeletal muscle biology and the emergence of targeted therapies, the successful translation of these innovations into clinical practice remains constrained by a range of scientific, regulatory, and operational challenges. Addressing these obstacles requires not only novel tools and interdisciplinary collaboration but also a systems-level rethinking of how skeletal muscle disorders are studied, treated, and monitored in diverse patient populations. One of the foremost barriers to clinical translation is the heterogeneity of muscle disorders themselves. Etiologies such as sarcopenia, cachexia, disuse atrophy, and genetic myopathies differ markedly in their molecular underpinnings, progression, and response to intervention. Even within a single diagnostic category, variability in age, sex, comorbidities, nutritional status, and physical activity levels complicates the generalizability of trial results and the development of standardized treatment protocols. This biological and phenotypic diversity necessitates the development of stratified or personalized approaches, where interventions are tailored based on molecular or functional profiles. Ethical and regulatory hurdles further limit the pace of innovation. Advanced therapies such as gene editing, iPSC-based interventions, and viral vector delivery systems raise unresolved questions regarding long-term safety, off-target effects, germline transmission, and equitable access. Regulatory agencies require rigorous demonstration of safety and efficacy, yet many preclinical models lack sufficient predictive validity, and long-term human data are scarce. Meanwhile, traditional randomized controlled trials (RCTs) may not be well-suited to test dynamic, adaptive interventions, such as machine-learning–driven neuromodulatory platforms or real-time nutritional algorithms, which require novel trial designs and regulatory flexibility.

Emerging multi-omics technologies offer a potential solution to this complexity by enabling a granular, systems-level view of muscle biology. Integration of transcriptomic, proteomic, metabolomic, and epigenomic data allows for the identification of novel disease endotypes, predictive biomarkers, and therapy-responsive subpopulations [83]. These insights can facilitate the discovery of new drug targets, inform patient selection criteria, and provide surrogate endpoints for early-phase trials. Coupled with advances in bioinformatics and AI, these data-rich platforms support the transition toward precision rehabilitation, a model in which therapy is dynamically adapted to a patient's molecular and functional phenotype. Platforms such as EcoAI, which integrate wearable biosensors, real-time symptom tracking, adaptive electrical stimulation, and cloud-based analytics, exemplify the future of AI-driven muscle therapeutics. These systems allow for continuous monitoring of therapy response, early identification of treatment failure, and real-time optimization of stimulation parameters or behavioral interventions. As machine learning models evolve, they may enable predictive forecasting of muscle deterioration or recovery trajectories, further personalizing care delivery. The convergence of digital therapeutics with regenerative medicine, gene therapy, and pharmacologic interventions holds promise for multimodal, synergistic treatment paradigms. Future clinical applications of muscle therapeutics extend beyond traditional indications and may play pivotal roles in aging populations, spaceflight-induced deconditioning, and critical illness recovery. In geriatric medicine, maintaining muscle integrity is essential for preserving autonomy, reducing fall risk, and preventing hospitalization. In the context of long-duration space missions, microgravity-induced muscle atrophy poses a serious challenge to astronaut health and mission success, necessitating the development of countermeasures that include closed-loop NMES systems, nutritional protocols, and hormonal modulators. Similarly, in intensive care and post-acute care settings, early mobilization strategies supported by remote monitoring and muscle-preserving interventions could accelerate functional recovery and reduce the burden of long-term disability.

Advancing the field of muscle therapeutics will require more than mechanistic insight; it demands an infrastructure that supports interdisciplinary innovation, inclusive trial design, equitable access, and real-time data integration. A forward-looking framework must accommodate both high-tech innovations and scalable, low-barrier interventions that can be deployed across diverse global populations. The next generation of muscle-targeted therapies will not only reverse decline but may also enhance resilience, ushering in a new era of regenerative, adaptive, and preventive muscle health.

8. Conclusion

Skeletal muscle represents a biologically complex and clinically pivotal organ system whose function extends well beyond locomotion. As both a contractile and endocrine tissue, it plays an essential role in metabolic regulation, systemic inflammation, and recovery from illness or injury. Disruption of muscle homeostasis, whether through genetic mutations, disuse, systemic disease, or aging, initiates a cascade of molecular events that include impaired protein synthesis, satellite cell exhaustion, mitochondrial dysfunction, and dysregulated inflammatory signaling. Understanding the precise molecular mechanisms governing myogenesis, hypertrophy, atrophy, and regeneration has yielded a detailed framework from which to develop targeted therapies. Therapeutic strategies have evolved considerably in recent years, moving from empirical interventions to mechanistically informed, multimodal approaches. Pharmacologic agents such as myostatin inhibitors, SARMs, and anti-inflammatory compounds aim to modulate key anabolic and catabolic signaling pathways. Gene editing technologies like CRISPR/Cas9, alongside AAV-mediated gene transfer and stem cell–based therapies, offer unprecedented potential for durable correction of primary muscle pathologies.

DOI: http://dx.doi.org/10.5772/intechopen.1011812

Electrical and biophysical modalities, including next-generation neuromuscular stimulation systems, are increasingly supported by digital health platforms capable of personalizing therapy in real time. Nutritional and resistance-based interventions remain foundational but are now enhanced through remote patient monitoring, wearable sensors, and AI-guided feedback systems that drive adherence and precision. As skeletal muscle biology becomes more integrated with the fields of regenerative medicine, bioengineering, and digital therapeutics, the traditional boundaries between pharmacologic, mechanical, and cellular therapies are dissolving. The future of muscle therapeutics lies in the convergence of these disciplines, where interventions are not only molecularly targeted but also behaviorally adaptive, remotely monitored, and dynamically optimized. With continued advancements in multi-omics, machine learning, and personalized rehabilitation platforms, skeletal muscle is poised to become both a therapeutic target and a biomarker-rich window into broader systemic health.

Through this translational lens, the evolving science of skeletal muscle offers not only restoration of function but also the potential to extend vitality, resilience, and quality of life across the human lifespan.

Author details

Maja Green[1,2*] and Krishnan Chakravarthy[1,2]

1 Solaris Research Institute, San Diego, Delaware, USA

2 NXTSTIM Inc, San Diego, California, USA

*Address all correspondence to: maja.lovric@hotmail.com

References

[1] Argilés JM, Campos N, Lopez-Pedrosa JM, et al. Skeletal muscle regulates metabolism via interorgan crosstalk: Roles in health and disease. Journal of the American Medical Directors Association 2016;**17**:789–796

[2] Tieland M, Trouwborst I, Clark BC. Skeletal muscle performance and ageing. Journal of Cachexia, Sarcopenia and Muscle. 2018;**9**(1):3–19

[3] Pratesi A, Tarantini F, Di Bari M. Skeletal muscle: An endocrine organ. Clinical Cases in Mineral and Bone Metabolism. 2013;**10**(1):11–14

[4] Weijs PJ, Looijaard WG, Dekker IM, et al. Low skeletal muscle area is a risk factor for mortality in mechanically ventilated critically ill patients. Journal of Critical Care 2014;**18**:R12

[5] Endo T. Molecular mechanisms of skeletal muscle development, regeneration, and osteogenic conversion. Bone. 2015;**80**:2–13

[6] Aweida D, Cohen S.Breakdown of filamentous myofibrils by the UPS–step by step. Biomolecules. 2021;**11**(1):110

[7] Feng L-T, Chen Z-N, Bian H.Skeletal muscle: Molecular structure, myogenesis, biological functions, and diseases. MedComm. 2024;**5**(7):e649

[8] Akinrodoye MA, Lui F. Neuroanatomy Somatic Nervous System. In StatPearls. Treasure Island (FL): StatPearls PublishingCopyright © 2025, StatPearls Publishing LLC; 2025

[9] Braun T, Gautel M.Transcriptional mechanisms regulating skeletal muscle differentiation, growth and homeostasis. Nature Reviews Molecular Cell Biology. 2011;**12**(6):349–361

[10] Lovric MM, Hawkins CJ.TRAIL treatment provokes mutations in surviving cells. Oncogene. 2010;**29** (36):5048–5060

[11] Hernández-Hernández JM, García-González EG, Brun CE, et al. The myogenic regulatory factors, determinants of muscle development, cell identity and regeneration. Seminars in Cell and Developmental Biology. 2017;**72**:10–18

[12] Meloux A, Rochette L, Maza M, et al. Growth differentiation factor-8 (GDF8)/myostatin is a predictor of troponin I peak and a marker of clinical severity after acute myocardial infarction. Journal of Clinical Medicine. 2019;**9**:116

[13] Mossahebi-Mohammadi M, Quan M, Zhang J-S, Li X, et al. FGF signaling pathway: A key regulator of stem cell pluripotency. Frontiers in Cell and Developmental Biology. 2020;**8**:79

[14] Park Y-J, Yoo S-A, Kim M. et al. The role of calcium–calcineurin–NFAT signaling pathway in health and autoimmune diseases. Frontiers in Immunology. 2020;**11**:195

[15] Ferraro E, Giammarioli AM, Chiandotto S, et al. Exercise-induced skeletal muscle remodeling and metabolic adaptation: Redox signaling and role of autophagy. Antioxidants & Redox Signaling 2014;**21**:154–176

[16] Flück M.Functional, structural and molecular plasticity of mammalian skeletal muscle in response to exercise

stimuli. Journal of Experimental Biology. 2006;**209**(12):2239–2248

[17] Chen W, Zhao H, Li Y. Mitochondrial dynamics in health and disease: Mechanisms and potential targets. Signal Transduction and Targeted Therapy. 2023;**8**(1):333

[18] Schrepfer E, Scorrano L. Mitofusins, from mitochondria to metabolism. Molecular Cell. 2016;**61** (5):683–694

[19] Ray PD, Huang B-W, Tsuji Y. Reactive oxygen species (ROS) homeostasis and redox regulation in cellular signaling. Cellular Signalling. 2012;**24**(5):981–990

[20] Juan CA, Pérez de la Lastra JM, Plou FJ, et al. The chemistry of reactive oxygen species (ROS) revisited: Outlining their role in biological macromolecules (DNA, lipids and proteins) and induced pathologies. International Journal of Molecular Sciences. 2021;**22**:4642

[21] Matsumura K, Teranaka W, Matsumoto H, et al. Loss of skeletal muscle mass predicts cardiac death in heart failure with a preserved ejection fraction but not heart failure with a reduced ejection fraction. ESC Heart Failure. 2020;**7**:4100–4107

[22] Xing Y-Q, Li A, Yang Y, Li -X-X, Zhang L-N, Guo H-C, et al. The regulation of FOXO1 and its role in disease progression. Life Sciences. 2018;**193**:124–131

[23] Bodine SC, Stitt TN, Gonzalez M, et al. Akt/mTOR pathway is a crucial regulator of skeletal muscle hypertrophy and can prevent muscle atrophy in vivo. Nature Cell Biology. 2001;**3**:1014–1019

[24] Magaye RR, Savira F, Hua Y, et al. Attenuating PI3K/Akt-mTOR pathway reduces dihydrosphingosine 1 phosphate mediated collagen synthesis and hypertrophy in primary cardiac cells. The International Journal of Biochemistry & Cell Biology. 2021;**134**:105952

[25] Sandri M.Protein breakdown in muscle wasting: Role of autophagy-lysosome and ubiquitin-proteasome. The International Journal of Biochemistry & Cell Biology. 2013;**45** (10):2121–2129

[26] Arends J, Bachmann P, Baracos V, et al. ESPEN guidelines on nutrition in cancer patients. Clinical Nutrition. 2017;**36**:11–48

[27] Gao Y, Arfat Y, Wang H, Goswami N, et al. Muscle atrophy induced by mechanical unloading: Mechanisms and potential countermeasures. Frontiers in Physiology. 2018;**9**:235

[28] Sartori R, Romanello V, Sandri M. Mechanisms of muscle atrophy and hypertrophy: Implications in health and disease. Nature Communications. 2021;**12**(1):330

[29] Larsson L, Degens H, Li M, et al. Sarcopenia: Aging-related loss of muscle mass and function. Physiological Reviews. 2019;**99**:427–511

[30] Rolland Y, Czerwinski S, van Kan GA, et al. Sarcopenia: Its assessment, etiology, pathogenesis, consequences and future perspectives. The Journal of Nutrition, Health & Aging. 2008;**12**:433–450

[31] Saini S, Kakati P, Singh K, et al Role of Inflammation in Tissue Regeneration and Repair. Tripathi A, Dwivedi A, Gupta S. editors Inflammation Resolution and Chronic Diseases. Singapore: Springer Nature Singapore 2024: 103–127.

[32] Forbes SJ, Rosenthal N.Preparing the ground for tissue regeneration: From mechanism to therapy. Nature Medicine. 2014;**20**(8):857–869

[33] Zheng X, Liu X, Guo Y, et al. Physical exercise and epigenetic modifications in skeletal muscle, brain, and heart. Epigenetics and Chromatin. 2025;**18**:12

[34] Massenet J, Gardner E, Chazaud B, et al. Epigenetic regulation of satellite cell fate during skeletal muscle regeneration. Skeletal Muscle. 2021;**11**:4

[35] Dai Y, Wang YM, Zhang WR, et al. The role of microRNA-1 and microRNA-206 in the proliferation and differentiation of bovine skeletal muscle satellite cells. In Vitro Cellular & Developmental Biology – Animal 2016;**52**:27–34

[36] Horak M, Novak J, Bienertova-Vasku J.Muscle-specific microRNAs in skeletal muscle development. Developmental Biology. 2016;**410**(1): 1–13

[37] Koutsoulidou A, Mastroyiannopoulos NP, Furling D, et al. Expression of miR-1, miR-133a, miR-133b and miR-206 increases during development of human skeletal muscle. BMC Developmental Biology. 2011;**11**:34

[38] Yin L, Li N, Jia W, et al. Skeletal muscle atrophy: From mechanisms to treatments. Pharmacological Research. 2021;**172**:105807

[39] Ji Y, Li M, Chang M, et al. Inflammation: Roles in skeletal muscle atrophy. Antioxidants (Basel). 2022;**11** (9):1350

[40] Cacciatore S, Calvani R, Esposito I, et al. Emerging targets and treatments for sarcopenia: A narrative review. Nutrients. 2024;**16**: 3271. DOI: 10.3390/nu16193271

[41] Passey SL, Hansen MJ, Bozinovski S, et al. Emerging therapies for the treatment of skeletal muscle wasting in chronic obstructive pulmonary disease. Pharmacology and Therapeutics. 2016;**166**:56–70

[42] Meriggioli MN, Roubenoff R. Prospect for pharmacological therapies to treat skeletal muscle dysfunction. Calcified Tissue International. 2015;**96** (3):234–242

[43] Kicman AT.Pharmacology of anabolic steroids. British Journal of Pharmacology. 2008;**154**(3):502–521

[44] West DW, Phillips SM.Anabolic processes in human skeletal muscle: Restoring the identities of growth hormone and testosterone. The Physician and Sportsmedicine. 2010;**38** (3):97–104

[45] Rizk J, Sahu R, Duteil D. An overview on androgen-mediated actions in skeletal muscle and adipose tissue. Steroids. 2023;**199**:109306

[46] Solomon ZJ, Mirabal JR, Mazur DJ, et al. Selective androgen receptor modulators: Current knowledge and clinical applications. Sexual Medicine Reviews. 2019;7:84–94

[47] Crawford J, Prado CMM, Johnston MA, et al. Study design and rationale for the phase 3 clinical development program of enobosarm, a selective androgen receptor modulator, for the prevention and treatment of muscle wasting in cancer patients (POWER trials). Current Oncology Reports. 2016;**18**:37

[48] Lee S-J. Targeting the myostatin signaling pathway to treat muscle loss and metabolic dysfunction. Journal of Clinical Investigation. 2021;**131**(9):e148372

[49] Hoogaars WMH, Jaspers JRT. Past, present, and future perspective of targeting myostatin and related signaling pathways to counteract muscle atrophy. Advances in Experimental Medicine and Biology. 2018;**1088**:153–206

[50] Wetzlich B, Nyakundi BB, Yang J. Therapeutic applications and challenges in myostatin inhibition for enhanced skeletal muscle mass and functions. Molecular & Cellular Biochemistry. 2025;**480**(3):1535–1553

[51] Cheng Y, Lin S, Cao Z, et al. The role of chronic low-grade inflammation in the development of sarcopenia: Advances in molecular mechanisms. International Immunopharmacology. 2025;**147**:114056

[52] Li X, Li C, Zhang W, et al. Inflammation and aging: Signaling pathways and intervention therapies. Signal Transduction and Targeted Therapy. 2023;**8**:239

[53] Giles JT, Sattar N, Gabriel S, et al. Cardiovascular safety of tocilizumab versus etanercept in rheumatoid arthritis: A randomized controlled trial. Arthritis & Rheumatology. 2020;**72**:31–40

[54] Cazzola M, Calzetta L, Matera MG. β 2 -adrenoceptor agonists: Current and future direction. British Journal of Pharmacology. 2011;**163**(1):4–17

[55] Marei HE.Stem cell therapy: A revolutionary cure or a pandora's box. Stem Cell Research and Therapy. 2025;**16**(1):255

[56] Aljabali AAA, El-Tanani M, Tambuwala MM. Principles of CRISPR-Cas9 technology: Advancements in genome editing and emerging trends in drug delivery. Journal of Drug Delivery Science & Technology. 2024;**92**:105338

[57] Xu Y, Li Z. CRISPR-Cas systems: Overview, innovations and applications in human disease research and gene therapy. Computational and Structural Biotechnology Journal. 2020;**18**:2401–2415

[58] Ebner N, Anker SD, von Haehling S.Recent developments in the field of cachexia, sarcopenia, and muscle wasting: Highlights from the 11th Cachexia Conference. Journal of Cachexia, Sarcopenia and Muscle. 2019;**10**(1):218–225

[59] Relaix F, Bencze M, Borok MJ, et al. Perspectives on skeletal muscle stem cells. Nature Communications. 2021;**12**:692

[60] Tedesco FS, Dellavalle A, Diaz-Manera J, et al. Repairing skeletal muscle: Regenerative potential of skeletal muscle stem cells. Journal of Clinical Investigation. 2010;**120**:11–19

[61] Rinaldi F, Perlingeiro RCR.Stem cells for skeletal muscle regeneration: Therapeutic potential and roadblocks. Translational Research. 2014;**163**(4):409–417

[62] Patel PM, Green M, Tram J, et al. Beyond the pain management clinic: The role of AI-integrated remote patient monitoring in chronic disease management – a narrative review. Journal of Pain Research. 2024;**17**:4223–4237

[63] Green M, Hayley A, Gunnersen JM, et al. Transforming chronic pain management: Integrating neuromodulation with advanced technologies to tackle cognitive dysfunction – a narrative review. Journal of Pain Research. 2025;**18**:2497–2507

[64] Patel P, Green M, Tram J, et al. Latest advancements in transcutaneous electrical nerve stimulation (TENS) and electronic muscle stimulation (EMS): Revisiting an established therapy with new possibilities. Journal of Pain Research. 2025;**18**:137–153

[65] Liu M, Luo J, Zhou J, et al. Intervention effect of neuromuscular electrical stimulation on ICU acquired weakness: A meta-analysis. International Journal of Nursing Sciences. 2020;**7**:228–237

[66] Marquez-Chin C, Popovic MR. Functional electrical stimulation therapy for restoration of motor function after spinal cord injury and stroke: A review. BioMedical Engineering On Line. 2020;**19**(1):34

[67] Duan H, Chen S, Mai X, et al. Low-intensity pulsed ultrasound (LIPUS) promotes skeletal muscle regeneration by regulating PGC-1α/AMPK/GLUT4 pathways in satellite cells/myoblasts. Cellular Signalling. 2024;**117**:111097

[68] Shanmugasundaram Prema S, Ganapathy D, Shanmugamprema D. Prehabilitation strategies: Enhancing surgical resilience with a focus on nutritional optimization and multimodal interventions. Advances in Nutrition. 2025;**16**(4):100392

[69] Carbone JW, Pasiakos SM.Dietary protein and muscle mass: Translating science to application and health benefit. Nutrients. 2019;**11**(5):1136

[70] Damas F, Phillips S, Vechin FC, et al. A review of resistance training-induced changes in skeletal muscle protein synthesis and their contribution to hypertrophy. Sports Medicine (Auckland, N.Z.). 2015;**45**:801–807

[71] Luo L, Lu A-M, Wang Y, et al. Chronic resistance training activates autophagy and reduces apoptosis of muscle cells by modulating IGF-1 and its receptors, Akt/mTOR and Akt/FOXO3a signaling in aged rats. Experimental Gerontology. 2013;**48**:427–436

[72] Chen Y, Li W, Hussam AS, et al. Transforming health and wellness: Exploring the captivating convergence of rehabilitation, exercise, and cutting-edge health gadgets in the rapidly evolving tech-driven world. Technology in Society. 2025;**81**:102808

[73] Campbell WW Nutrition for sport and exercise: 2nd ed, edited by Jacqueline R Berning and Suzanne Nelson Steen. The American Journal of Clinical Nutrition. 1999; **69**(3): 578–579. 297 pages, hardcover, $52.00 Aspen Publishers, Inc, Gaithersburg, MD.

[74] Mukherjee P, Roy S, Ghosh D, et al. Role of animal models in biomedical research: A review. Laboratory Animal Research. 2022;**38**:18

[75] Domínguez-Oliva A, Hernández-Ávalos I, Martínez-Burnes J, et al. The importance of animal models in

biomedical research: Current insights and applications. Animals. 2023;**13**: 1223. DOI: 10.3390/ani13071223

[76] Kornegay JN, Bogan JR, Bogan DJ, et al. Golden retriever muscular dystrophy (GRMD): Developing and maintaining a colony and physiological functional measurements. Methods Molecule Biology. 2011;**709**:105–123

[77] Goldstein RA.Skeletal muscle injury biomarkers: Assay qualification efforts and translation to the clinic. Toxicologic Pathology. 2017;**45**(7): 943–951

[78] Huang W, Sowa G.Biomarker development for musculoskeletal diseases. Physical Medicine and Rehabilitation. 2011;**3**(6S):S39–S44

[79] Muollo V, D'Emanuele S, Ghiotto L, et al. Evaluating handgrip strength and functional tests as indicators of gait speed in older females. Frontiers in Sports and Active Living. 2025;7:1497546

[80] Mendell JR, Muntoni F, McDonald CM, et al. AAV gene therapy for Duchenne muscular dystrophy: The EMBARK phase 3 randomized trial. Nature Medicine. 2025;**31**:332–341

[81] Kanbay M, Siriopol D, Copur S, et al. Effect of bimagrumab on body composition: A systematic review and meta-analysis. Aging Clinical and Experimental Research. 2024;**36**:185

[82] Christiansen AR, Lipshultz LI, Hotaling JM, et al. Selective androgen receptor modulators: The future of androgen therapy? Translational Andrology and Urology. 2020;**9**: S135–s148

[83] Baysoy A, Bai Z, Satija R, et al. The technological landscape and applications of single-cell multi-omics. Nature Reviews Molecular Cell Biology. 2023;**24**:695–713

Chapter 2

Muscle-Nerve-Immune Crosschats: Troika Theorization

Debprasad Dutta

Abstract

The muscle-nerve-immune "troika" – a dynamic, interdependent network critical for skeletal muscle health, disease, and repair. Moving beyond traditional siloed views, troika theory postulates that muscle fibers, neurons, and immunocytes, along with the molecular milieu resulting from their cross-talk, enable homeostasis, attune responses to injury, and facilitate tissue regeneration. Key mechanisms of cross-talk among the muscular, nervous, and immune systems are responsible for health, and the perturbation of troika homeostasis induces diseases. The troika model encompasses major biomolecules and signaling pathways governing tri-systemic interaction. Exemplary multisystem pathologies driven by troika dysregulation, including muscular dystrophies, autoimmune neuropathies, and age-related neurodegenerative conditions, are discussed. The translational and clinical potential of the troika rubric lies in interactional signatures that can serve as early risk prediction biomarkers, diagnostic tools, and novel therapeutic targets. The integration of multiomics, machine learning, and high-throughput analytical modalities accentuates the promise of precision medicine in disorders with troika involvement.

Keywords: myoneural, neuroimmune, neuromuscular, multisystem diseases, interorgan communication, multiomics integration

1. Introduction

1.1 Overview of skeletal muscle physiology

Skeletal muscle is a fundamental tissue of the human body, accounting for approximately 30–40% of total body mass and playing a pivotal role in voluntary movement, locomotion, posture, and overall metabolic homeostasis [1]. Skeletal muscle fibers are highly specialized, multinucleated cells characterized by a complex arrangement of contractile proteins organized into sarcomeres, which underpin their ability to generate force and movement through excitation-contraction coupling. Beyond its mechanical functions, skeletal muscle serves as a metabolic organ involved in glucose uptake, fatty acid oxidation, and secretion of myokines (signaling molecules released by muscle that influence systemic physiology). The cellular architecture of skeletal muscle features satellite cells, a population of muscle stem cells (MuSC) that are indispensable for muscle growth, repair, and regeneration following injury or stress. The excitation-contraction process depends on neuromuscular signaling, calcium dynamics within the sarcoplasmic reticulum, and meticulous coordination of molecular events to convert chemical energy into mechanical force.

Skeletal muscle physiology reflects a dynamic balance between anabolic and catabolic processes, adapting to various physiological demands such as exercise, aging, and pathological insults. Its performance and integrity hinge on the interplay between cellular metabolism, structural organization, and intricate signaling pathways that regulate protein turnover, mitochondrial function, and cellular communication. Consequently, disruptions in any of these components may lead to muscle weakness, wasting, and impaired function, underscoring the necessity of an integrative understanding of muscle biology.

1.2 Traditional views on muscle, nerve, and immune systems

Stereotypically, skeletal muscle, nervous system, and immune system research have been conducted within distinct disciplinary silos, focusing on isolated organ functions and cellular roles. Muscle biology has primarily centered on contractile mechanics, fiber-type composition, and metabolic capacity. Neurobiology has detailed the roles of motor and sensory neurons in controlling muscle activity and conveying proprioceptive information essential for coordinated movement. Immunology has examined immune surveillance and responses primarily within the context of infection and inflammation, often detached from muscular and neuronal considerations.

Although these domains have developed robust knowledge individually, this stereotypical siloed approach has limited comprehension of the systemic interdependencies critical for maintaining musculoskeletal health [2–4]. For instance, the nervous system not only initiates muscle contraction but also modulates trophic support via neurotrophic factors. Similarly, immune cells, long regarded merely as responders to injury or infection, are now recognized as active regulators of muscle remodeling and neural plasticity. Emerging cross-disciplinary research reveals that interactions among muscle, nerve, and immune cells occur continuously during physiological states and are vital for tissue adaptation, repair, and stress response.

Failure to appreciate these interactive networks contributes to incomplete models of disease, where musculoskeletal conditions are often accompanied by neuroimmune alterations. Conditions such as autoimmune myopathies, neuropathies, and degenerative muscle diseases manifest clinical features that reflect the integrated dysfunction among these systems rather than isolated pathology. Thus, there has been a paradigm shift toward viewing muscle, nerve, and immune functions as a triadic system whose cross-talk underlies both health and disease processes.

1.3 Rationale for the troika rubric

The troika rubric is proposed as a conceptual model to encapsulate the dynamic and reciprocal interactions among skeletal muscle fibers, peripheral nerve components, and immune cell populations, forming an interdependent and

DOI: http://dx.doi.org/10.5772/intechopen.1014066

regulatory triad. This model transcends reductionist approaches by emphasizing the systemic nature of muscle health and pathophysiology. It provides a scaffold to integrate molecular, cellular, and physiological information from traditionally separate fields into a coherent narrative explaining how homeostasis is achieved and disrupted.

Central to the troika rubric is the recognition that each node in the triad continuously communicates through direct cell-cell contact, paracrine signaling, and systemic mediators such as cytokines, chemokines, neurotransmitters, and growth factors. This communication modulates cellular behaviors, including survival, differentiation, inflammation, and regeneration, which collectively determine tissue outcomes. Importantly, signals originating from muscle fibers influence neural and immune responses; reciprocally, neuronal input modulates immune activity and muscle function, while immune cells regulate neurogenic and myogenic processes.

This integrative perspective also rationalizes the occurrence of complex multisystem disorders, where perturbations in one domain propagate deleterious effects across the network. The troika rubric offers a strategic approach to identify novel biomarkers, elucidate disease mechanisms, and develop multifaceted therapies targeting this coordinated system rather than isolated components. It serves as a foundation for future research aimed at uncovering the multidimensional interactions crucial for musculoskeletal health, with implications spanning from basic biology to clinical translation. A visual rendition of the troika theory is provided in **Figure 1**.

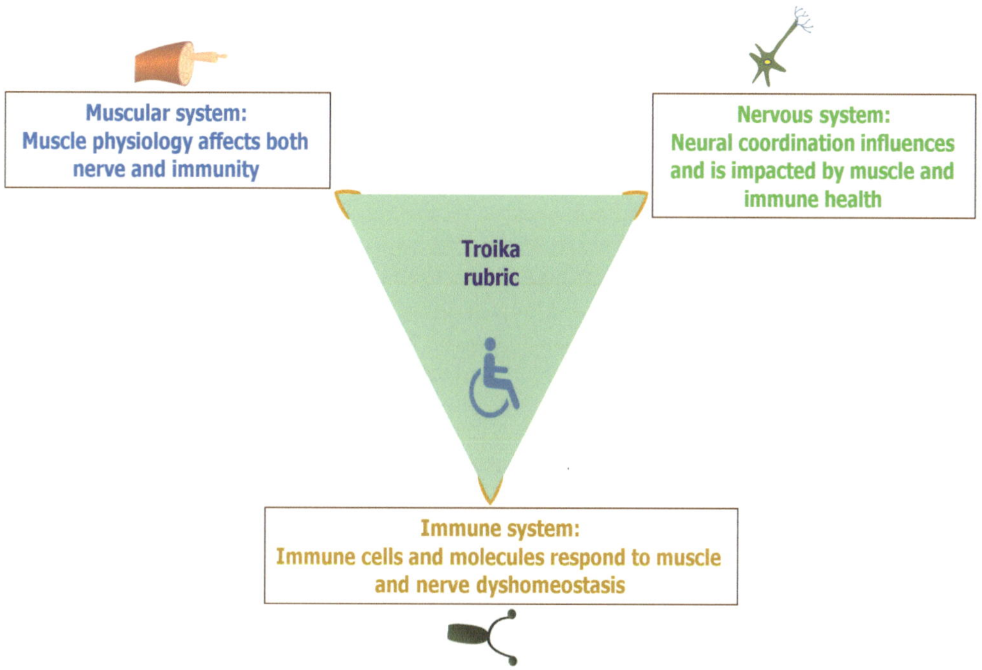

Figure 1.
Foundations of "troika" theory: schematic representation of the muscle-nerve-immune "troika" rubric. The diagram illustrates that the muscle-nerve-immune "troika" is a complex, dynamic, and interdependent network essential for skeletal muscle health, disease progression, and tissue repair. The central triangle evinces integrated three-way cross-talks essential for homeostasis and disease response.

2. Structural basis of muscle-nerve-immune interaction

2.1 Anatomy of skeletal muscle and innervation

Skeletal muscle is composed of bundles of muscle fibers, each fiber being a multinucleated cell packed with contractile proteins organized into sarcomeres. Motor neurons innervate muscle fibers at specialized synapses to initiate contraction, while sensory neurons provide feedback on muscle status. Sympathetic nerves regulate blood flow and metabolism, contributing to muscle function. Effective neuromuscular communication is essential for movement and muscle health [5].

2.2 Cellular components of the immune system in muscle

Muscle hosts resident immune cells such as macrophages, dendritic cells, and T cells, distributed mainly in connective tissue layers. These cells respond to injury by clearing debris, modulating inflammation, and supporting muscle repair through growth factor secretion [6]. Immune cells closely interact with MuSCs, orchestrating regeneration and remodeling in a dynamic and context-dependent manner.

2.3 Neuromuscular junction: Interface of nerve and muscle

The neuromuscular junction (NMJ) is a highly specialized synapse where motor neurons transmit signals to muscle fibers by releasing acetylcholine, which binds to receptors on the muscle membrane to trigger contraction. Supported by Schwann cells, the NMJ is a critical site of neuromuscular communication and plasticity [7, 8]. Its integrity is vital for muscle function and is sensitive to neuroimmune modulation.

2.4 Connective tissue and vascular networks facilitating crosschats

The extracellular matrix and connective tissue layers (endomysium, perimysium, epimysium) provide structural scaffolding and support for force transmission. Embedded vascular networks facilitate the delivery of nutrients and the trafficking of immune cells. Pericytes and endothelial cells within vessels contribute signaling cues that integrate muscle, nerve, and immune functions, forming a microenvironment essential for tissue homeostasis and repair [9].

3. Molecular mediators of troika crosschats

3.1 Cytokines and chemokines in muscle-immune signaling

Cytokines and chemokines are key signaling proteins mediating communication between muscle and immune cells. Muscle fibers and infiltrating immune cells secrete proinflammatory cytokines like TNF-α and chemokines such as CCL2 which recruit immune cells to injury sites and regulate inflammation [10]. These molecules orchestrate immune cell activation and muscle regeneration by modulating satellite cell behavior, debris clearance, and extracellular matrix remodeling. IL-6 has both proinflammatory and antiinflammatory effects in muscle tissue, with its dual role depending on the specific signaling pathway activated and the overall physiological context, such as exercise or chronic disease [11]. During exercise, muscle-derived IL-6 primarily acts as an antiinflammatory agent and energy allocator by enhancing glucose uptake and lipolysis [12]. In contrast, when secreted by other cells in response to chronic inflammation, IL-6 can exacerbate muscle wasting (atrophy), and hindering IL-6/JAK/STAT3 signaling helps repair it [13]. The above-cited evidence suggests that balanced cytokine signaling is essential; dysregulation can exacerbate chronic inflammation and muscle pathology.

3.2 Neurochemicals affecting muscle and immune cells

Neurotransmitters, including acetylcholine and norepinephrine, influence both muscle contraction and immune cell activity within the muscle microenvironment. Neurotrophic factors like brain-derived neurotrophic factor (BDNF) and glial cell-derived neurotrophic factor (GDNF) support neuronal survival and promote muscle regeneration via innervation [14, 15]. These molecules mediate neuroimmune cross-talks by modulating inflammation and facilitating repair processes, highlighting their integral role in maintaining troika homeostasis.

3.3 Receptor systems and signal integration across cell types

Muscle, nerve, and immune cells express a variety of receptors – cytokine receptors, neurotransmitter receptors, and pattern recognition receptors, especially toll-like receptors (TLRs) – enabling them to detect and respond to cross-system signals [16, 17]. Signal integration through these receptors governs cellular behaviors such as proliferation, differentiation, and immune activation. The convergence of multiple signaling pathways facilitates coordinated responses essential for tissue homeostasis and effective repair.

3.4 Emerging molecular pathways and their functional roles

Evidence highlights crucial pathways involved in troika interactions, including NF-κB signaling, JAK-STAT pathways, interaction with gangliosides, and inflammasome activation [18]. These pathways control gene expression programs that regulate inflammation, cell survival, and regeneration across muscle, neural, and immune cells. Understanding these emerging mechanisms offers insights into troika dysregulation in diseases and presents opportunities for targeted therapeutic interventions.

4. Physiology of muscle-nerve-immune homeostasis

4.1 Mechanisms maintaining troika equilibrium in health

Homeostasis in the muscle-nerve-immune troika is maintained through tightly regulated cellular and molecular interactions that balance proinflammatory and antiinflammatory signals, neuromodulation, and tissue integrity. Immune surveillance by resident cells works alongside neurogenic control to modulate inflammatory responses and ensure effective muscle function [19]. Key pathways involve cytokine signaling, neuroimmune synapses, and metabolic regulation to sustain equilibrium and prevent pathological activation.

4.2 Adaptive responses to physiological stress and exercise

Exercise and physiological stress induce adaptive changes across the troika. Muscle contraction releases myokines that modulate immune responses and support neural plasticity [20]. Concurrently, sympathetic nervous system activation influences immune cell trafficking and cytokine production [21]. These coordinated adaptations enhance muscle endurance, promote antiinflammatory environments, and optimize repair mechanisms, highlighting the dynamic plasticity within the troika.

4.3 Repair and regeneration processes involving all three systems

Following injury, the troika collaborates to initiate repair through immune cell recruitment, satellite cell activation, and neuronal remodeling. Macrophages clear debris and secrete factors aiding satellite cell proliferation. Neural inputs modulate inflammation and support reinnervation, critical for muscle recovery [22]. Feedback between these systems governs the timely resolution of inflammation and restoration of tissue architecture.

4.4 Feedback loops and network controls

Complex feedback mechanisms and network controls involving cytokines, neurotransmitters, and growth factors regulate troika functions. Positive and negative feedback loops ensure precise modulation of immune activation, neuronal excitability, and muscle metabolism [23]. Dysregulated feedback can lead to chronic inflammation or degeneration, emphasizing the necessity of integrated control to maintain tissue health.

5. Pathophysiological consequences of troika dysregulation

5.1 Muscular dystrophies and secondary neuroimmune effects

Muscular dystrophies, such as Duchenne muscular systrophy (DMD), result from genetic mutations disrupting structural proteins like dystrophin, causing muscle fiber fragility and progressive degeneration [24]. Muscle damage leads to chronic inflammation, with infiltration of macrophages and T cells amplifying muscle wasting. Neuroimmune interactions exacerbate the pathology by affecting motor neuron function and NMJ integrity [25], highlighting the contribution of immune dysregulation to disease severity.

5.2 Autoimmune neuropathies and neuromuscular disorders

Autoimmune neuropathies involve an immune-mediated attack on peripheral nerves, impairing neuronal signaling to muscles. Disorders like Guillain-Barré syndrome (GBS) and myasthenia gravis exemplify how aberrant immune responses disrupt neuromuscular transmission, causing muscle weakness [26]. The inflammation-induced damage triggers secondary muscle pathology, underlining the intertwined pathogenesis of nerve and immune system dysfunction in neuromuscular diseases.

5.3 Age-related sarcopenia and neurodegeneration

Aging leads to sarcopenia, characterized by loss of muscle mass and strength, accompanied by a decline in motor neuron number and function [27]. Chronic low-grade inflammation, or "inflammaging," activates immune pathways that impair muscle regeneration and neural plasticity. Neuroimmune imbalance contributes to progressive neuromuscular degeneration, emphasizing the importance of troika homeostasis in healthy aging.

5.4 Inflammatory myopathies and chronic inflammation in muscle

Inflammatory myopathies, including polymyositis and dermatomyositis, are characterized by sustained immune cell infiltration and cytokine production within muscle [28]. Persistent inflammation damages muscle fibers, disrupts NMJs, and impairs repair processes. Dysregulated troika signaling drives chronic disease progression, causing profound muscle weakness and disability.

6. Translational and clinical implications

6.1 Biomarker discovery exploiting troika signatures

Biomarkers derived from integrated muscle-nerve-immune signaling profiles hold promise for the sensitive detection and monitoring of neuromuscular diseases. Proteomic, transcriptomic, and metabolomic studies reveal candidate molecules such as cytokines, muscle-specific proteins, and neurotrophic factors that reflect disease activity and progression. Multiomics and machine learning (ML) analyses enhance the identification of troika signatures that could enable precision diagnostics and prognostication.

6.2 Diagnostic platforms for early detection of neuromuscular diseases

Advances in high-throughput and minimally invasive technologies have facilitated the development of diagnostic platforms that leverage troika biomarkers for early detection. Blood-based assays, imaging modalities, and electrophysiological tools capture neuropathic, myopathic, and immune dysfunction markers. Integration of these modalities promises earlier diagnosis, patient stratification, and personalized treatment planning.

6.3 Therapeutic targets in muscle-nerve-immune interplay

Targeting key molecular mediators within the troika offers novel therapeutic avenues. Modulating inflammatory cytokines, enhancing neurotrophic support, and stabilizing NMJs are strategies under investigation. Therapies aiming at restoring troika homeostasis have the potential to slow disease progression, improve muscle function, and mitigate secondary neuroimmune damage.

7. Future directions and research needs

7.1 Bridging clinical and experimental research

Advancing troika understanding requires integrative approaches that connect clinical observations with mechanistic experimental studies. Longitudinal patient cohorts, combined with animal models and cellular systems, can validate hypotheses and identify causative pathways [29–31]. Multidisciplinary collaborations between neurologists, immunologists, and muscle biologists are essential to translate findings into clinical practice.

7.2 Unresolved questions and hypotheses

Key gaps remain regarding the precise molecular mechanisms governing muscle-nerve-immune interactions, particularly in chronic and complex diseases. Questions include how the temporal dynamics of troika signaling influence disease onset, the role of novel cell types, and interactions with systemic factors like metabolism and microbiota. Addressing these unknowns will deepen the conceptual framework and therapeutic potential.

7.3 Technologies poised to advance troika research

Emerging technologies such as connectomics integration [32, 33] are poised to progress troika research. These tools enable high-resolution mapping of cellular interactions and pathways in health and disease, facilitating the identification of novel biomarkers and targets. Integration of multiscale data will provide holistic insights.

7.4 Potential for precision medicine and personalized therapies

Understanding patient-specific troika signatures opens avenues for precision medicine, enabling tailored interventions based on unique molecular, cellular, and systemic profiles. Personalized therapies targeting multiple troika nodes promise improved efficacy and reduced side effects. Future clinical trials incorporating troika-based biomarkers will refine patient stratification and treatment optimization.

Acronyms

BDNF	Brain-Derived Neurotrophic Factor
CCL2	Chemokine (C-C motif) Ligand 2
DMD	Duchenne Muscular Dystrophy
ECM	Extracellular Matrix
GBS	Guillain-Barré syndrome

IL-6	Interleukin-6
JAK	Janus Kinase
ML	Machine Learning
MuSC	Muscle Stem Cell
NF-κB	Nuclear Factor kappa B
NMJ	Neuromuscular Junction
STAT	Signal Transducer and Activator of Transcription
TNF-α	Tumor Necrosis Factor-alpha
TLR	Toll-Like Receptor
GDNF	Glial Cell-Derived Neurotrophic Factor

Author details

Debprasad Dutta
Mazumdar Shaw Center for Translational Research (MSCTR), Mazumdar Shaw Medical Foundation (MSMF), Narayana Health City, Bengaluru, India

*Address all correspondence to: debdutta.bio@gmail.com; debprasad@ms-mf.org

References

[1] Wong JPH, Ng YK, Kjærgaard J, Blazev R, Deshmukh AS, Parker BL. Skeletal muscle proteomics: Considerations and opportunities. NPJ Metabolic Health and Disease. 2025;**3**(1):1–10. DOI: 10.1038/S44324-025-00073-2

[2] Chu C, Artis D, Chiu IM. Neuro-immune Interactions in the Tissues. Immunity. 2020;**52**(3):464–474. DOI: 10.1016/J.IMMUNI.2020.02.017

[3] Nelke C, Dziewas R, Minnerup J, Meuth SG, Ruck T. Skeletal muscle as potential central link between sarcopenia and immune senescence. EBioMedicine. 2019;**49**:381–388. DOI: 10.1016/j.ebiom.2019.10.034

[4] Lømo T. "Nerve–muscle interactions". Handbook of Clinical Neurophysiology. 2003;**2**(C):47–65. DOI: 10.1016/S1567-4231(09)70114-5

[5] Pasnoor M, Dimachkie MM. Approach to Muscle and Neuromuscular Junction Disorders. Continuum (N Y). 2019;**25**(6):1536–1563. DOI: 10.1212/CON.0000000000000799

[6] Tidball JG. Regulation of muscle growth and regeneration by the immune system. Nature Reviews Immunology. 2017;**17**(3):165–178. DOI: 10.1038/nri.2016.150

[7] Barik A, Li L, Sathyamurthy A, Xiong WC, Mei L. Schwann Cells in Neuromuscular Junction Formation and Maintenance. The Journal of Neuroscience. 2016;**36**(38):9770. DOI: 10.1523/JNEUROSCI.0174-16.2016

[8] Auld DS, Robitaille R. Perisynaptic Schwann cells at the neuromuscular junction: Nerve- and activity-dependent contributions to synaptic efficacy, plasticity, and reinnervation. The Neuroscientist. 2003;**9**(2):144–157. DOI: 10.1177/1073858403252229

[9] Birbrair A, Zhang T, Wang ZM, Messi ML, Mintz A, Delbono O. Pericytes at the intersection between tissue regeneration and pathology. Clinical Science. 2015;**128**(2):81. DOI: 10.1042/CS20140278

[10] Waldemer-Streyer RJ, Kim D, Chen J. Muscle cell-derived cytokines in skeletal muscle regeneration. The FEBS Journal. 2022;**289**(21):6463. DOI: 10.1111/FEBS.16372

[11] Muñoz-Cánoves P, Scheele C, Pedersen BK, Serrano AL. Interleukin-6 myokine signaling in skeletal muscle: A double-edged sword? The FEBS Journal. 2013;**280**(17):4131. DOI: 10.1111/FEBS.12338

[12] Kistner TM, Pedersen BK, Lieberman DE. Interleukin 6 as an energy allocator in muscle tissue. Nature Metabolism. 2022;**4**(2):170–179. DOI: 10.1038/s42255-022-00538-4

[13] Huang Z, et al. Inhibition of IL-6/JAK/STAT3 pathway rescues denervation-induced skeletal muscle atrophy. Annals of Translational Medicine. 2020;**8**(24):1681–1681. DOI: 10.21037/ATM-20-7269

[14] Allen SJ, Watson JJ, Shoemark DK, Barua NU, Patel NK. GDNF, NGF and BDNF as therapeutic options for neurodegeneration. Pharmacology & Therapeutics. 2013;**138**(2):155–175. DOI: 10.1016/J.PHARMTHERA.2013.01.004

[15] Erickson JT, Brosenitsch TA, Katz DM. Brain-Derived Neurotrophic

Factor and Glial Cell Line-Derived Neurotrophic Factor Are Required Simultaneously for Survival of Dopaminergic Primary Sensory Neurons In Vivo. The Journal of Neuroscience. 2001;**21**(2):581–589. DOI: 10.1523/JNEUROSCI.21-02-00581.2001

[16] Debnath M, et al. Evidence of altered Th17 pathway signatures in the cerebrospinal fluid of patients with Guillain Barré Syndrome. Journal of Clinical Neuroscience. 2020;**75**(3):176–180. DOI: 10.1016/j.jocn.2020.03.010

[17] Dutta D, et al. Variations within Toll-like receptor (TLR) and TLR signaling pathway-related genes and their synergistic effects on the risk of Guillain-Barré syndrome. Journal of the Peripheral Nervous System. 2022;**27**(2):131–143. DOI: 10.1111/JNS.12484

[18] Dutta D, et al. Impact of Antecedent Infections on the Antibodies against Gangliosides and Ganglioside Complexes in Guillain-Barré Syndrome: A Correlative Study. Annals of Indian Academy of Neurology. 2022;**25**(3):401. DOI: 10.4103/AIAN.AIAN_121_22

[19] Xiang Y, Dai J, Xu L, Li X, Jiang J, Xu J. Research progress in immune microenvironment regulation of muscle atrophy induced by peripheral nerve injury. Life Sciences. 2021;**287**:120117. DOI: 10.1016/J.LFS.2021.120117

[20] Bay ML, Pedersen BK. Muscle-Organ Crosstalk: Focus on Immunometabolism. Frontiers in Physiology. 2020;**11**:567881. DOI: 10.3389/FPHYS.2020.567881

[21] Hodo TW, de Aquino MTP, Shimamoto A, Shanker A. Critical Neurotransmitters in the Neuroimmune Network. Frontiers in Immunology. 2020;**11**:1869. DOI: 10.3389/FIMMU.2020.01869

[22] Golshadi M, et al. Delay modulates the immune response to nerve repair. Npj Regenerative Medicine. 2023;**8**(1):1–13. DOI: 10.1038/s41536-023-00285-4

[23] Rahman A, Tiwari A, Narula J, Hickling T. Importance of Feedback and Feedforward Loops to Adaptive Immune Response Modeling. CPT: Pharmacometrics & Systems Pharmacology. 2018;**7**(10):621. DOI: 10.1002/PSP4.12352

[24] Lovering RM, Iyer SR, Edwards B, Davies KE. Alterations of neuromuscular junctions in Duchenne muscular dystrophy. Neuroscience Letters. 2020;**737**:135304. DOI: 10.1016/J.NEULET.2020.135304

[25] Aguilera M, Melgar S, Aguilera M, Melgar S. Microbial Neuro-Immune Interactions and the Pathophysiology of IBD. New Insights into Inflammatory Bowel Disease. 2016. DOI: 10.5772/64832

[26] Debnath M, et al. Comprehensive cytokine profiling provides evidence for a multi-lineage Th responses in Guillain Barré Syndrome. Cytokine. 2018;**110**(10):58–62. DOI: 10.1016/j.cyto.2018.04.026

[27] Wang N, et al. Aging induces sarcopenia by disrupting the crosstalk between the skeletal muscle microenvironment and myofibers. Journal of Advanced Research. 2025. DOI: 10.1016/J.JARE.2025.07.004

[28] Herbelet S, De Bleecker JL. Immune checkpoint failures in inflammatory myopathies: An overview. Autoimmunity Reviews.

2018;**17**(8):746–754. DOI: 10.1016/J.AUTREV.2018.01.026

[29] Melamud E, et al. The promise and reality of therapeutic discovery from large cohorts. Journal of Clinical Investigation. 2020;**130**(2):575. DOI: 10.1172/JCI129196

[30] Mele A, et al. In vivo longitudinal study of rodent skeletal muscle atrophy using ultrasonography. Scientific Reports. 2016;**6**(1):1–11. DOI: 10.1038/srep20061

[31] Ekert JE, et al. Recommended Guidelines for Developing, Qualifying, and Implementing Complex In Vitro Models (CIVMs) for Drug Discovery. SLAS Discovery. 2020;**25**(10):1174–1190. DOI: 10.1177/2472555220923332

[32] Dutta D. Crossroads Between Cognitive Connectomics and Sociomics: Synergies and Squabbles Amidst Two Omics. In: Reddy J, editor. Principles and Clinical Interventions in Social Cognition. ch. 17. Hershey, PA: IGI Global; 2024. p. 288–308. DOI: 10.4018/979-8-3693-1265-0.CH017

[33] Srinivasan R, Li Q, Zhou X, Lu J, Lichtman J, Wong STC. Reconstruction of the neuromuscular junction connectome. Bioinformatics. 2010;**26**(12):i64–i70. DOI: 10.1093/BIOINFORMATICS/BTQ179

Chapter 3

Metabolic Crosstalk Between Skeletal Muscle and Visceral Organs in Disease States

Zena Alani

Abstract

Skeletal muscle is increasingly recognised as a key component of the musculoskeletal system, in addition to being a vital, metabolically active endocrine organ that plays a central role in regulating whole-body energy homeostasis. Skeletal muscle communicates with various visceral tissues through the secretion of contraction-induced myokines, metabolites, and neurohormonal signals that establish complex bidirectional crosstalk with organs such as the liver, adipose tissue, pancreas, gut, and brain. This intricate network facilitates metabolic flexibility, enabling the body to adapt efficiently to different physiological and environmental demands. In health, this muscle–organ communication orchestrates processes essential for maintaining energy balance, glucose regulation, lipid metabolism, and overall metabolic health. However, in pathological conditions – including obesity, type 2 diabetes mellitus (T2DM), non-alcoholic fatty liver disease (NAFLD), cancer cachexia, chronic kidney disease (CKD), and critical illness – the normal regulatory mechanisms become disrupted, leading to maladaptive signalling. These perturbations contribute to systemic dysfunction and disease progression. This chapter synthesises the current understanding of muscle–organ crosstalk and highlights how these interactions are altered in disease. Clinical implications are emphasised, particularly in peri-operative and critical care contexts, including prehabilitation, nutrition strategies, and emerging therapies targeting the exercise–myokine axis.

Keywords: skeletal muscle, myokines, muscle–organ crosstalk, metabolic health, exercise–myokine axis

1. Introduction

Skeletal muscle forms approximately 40% of body mass and is the principal site of insulin-stimulated glucose disposal and fatty-acid oxidation [1]. Classically viewed as a contractile apparatus, muscle is now framed as an endocrine and paracrine organ that both responds to and shapes systemic metabolic cues through secreted factors and substrate flux [2, 3]. During fasting and stress, proteolysis supplies amino acids for hepatic gluconeogenesis, acute-phase protein synthesis, and immune cell proliferation [4]. During feeding and exercise, muscle serves as a sink for glucose and lipids, buffering glycaemia and circulating lipids while releasing myokines that signal to distant organs [2, 5].

Loss of muscle quantity and quality (sarcopenia and myosteatosis) is a powerful predictor of morbidity, mortality, length of hospital stay, and failure to discharge to home across medical and surgical specialties [4, 6]. Metabolic crosstalk can be conceptualised along three interlocking axes: (1) endocrine signalling via myokines and exerkines; (2) exchange of substrates such as glucose, lactate, amino acids, and lipids; and (3) autonomic and hormonal pathways that synchronise tissue responses to stress and exercise [2, 7]. Context dependence is central: interleukin-6 (IL-6) is insulin-sensitising when released transiently during exercise but contributes to insulin resistance and cachexia when chronically elevated in inflammation [8, 9].

Rather than linear pathways, muscle–organ interactions form an integrated adaptive network in which endocrine, metabolic, immune, and neural signals converge to coordinate systemic energy allocation [2, 10]. These networks are dynamically remodelled by nutrition, physical activity, inflammation, and disease burden, leading to shifts between anabolic resilience and catabolic vulnerability, particularly in peri-operative and chronic disease states [4, 6].

2. Mechanisms of muscle–organ crosstalk

2.1 Myokines and exerkines

Myokines are cytokines and peptide factors synthesised and released by myofibres in response to contraction, nutrient availability, hypoxia, or injury [2]. Examples include IL-6, IL-15, myostatin (GDF-8), irisin (cleaved from FNDC5), fibroblast growth factor-21 (FGF21), myonectin (CTRP15), apelin, follistatin-like 1 (FSTL1), and brain-derived neurotrophic factor (BDNF) [3]. Their actions span autocrine regulation of myogenesis and mitochondrial biogenesis; paracrine effects on vasculature and connective tissue; and endocrine control of hepatic glucose output, adipose lipolysis and browning, pancreatic insulin secretion, renal ammoniagenesis, and neuroplasticity [4].

Skeletal muscle also releases exosomes and other extracellular vesicles carrying micro-RNAs and proteins that reprogramme target tissues at a distance – an emerging layer of crosstalk termed exerkine signalling [5]. At the intracellular level, many myokines are regulated by contraction-induced calcium flux, AMPK activation, redox-sensitive transcription factors, and mitochondrial stress signalling, linking muscle energetic state directly to endocrine output [2, 3]. Increasing evidence also implicates muscle-derived extracellular vesicles as carriers of microRNAs that fine-tune insulin sensitivity, inflammatory tone, and mitochondrial function in target organs, adding a further layer of post-transcriptional regulation to inter-organ communication [5].

2.2 Substrate flux and the metabolic web

Substrate exchange couples muscle metabolism to visceral organ function. Post-prandially, insulin drives GLUT4 translocation to the sarcolemma, enabling muscle to dispose of the majority of circulating glucose and store it as glycogen [2]. During exercise, glycolytic fibres produce lactate which is shuttled to oxidative fibres or exported to the liver for gluconeogenesis (the Cori cycle) [6]. Amino acids, particularly alanine, glutamine, and the branched-chain amino acids (BCAAs), move from muscle to the liver and gut for gluconeogenesis and as fuel for immune cells [6]. Conversely, free fatty acids liberated from adipose tissue supply β-oxidation in working muscle; failure of this coupling leads to intramyocellular lipid accumulation and lipotoxic insulin resistance [6]. These substrate exchanges are tightly regulated by intracellular nutrient-sensing pathways, notably the AMPK–mTOR axis, which integrates energy availability with anabolic and catabolic decision-making [6]. Dysregulation of these pathways contributes to metabolic inflexibility, whereby muscle fails to appropriately switch between lipid and glucose oxidation, amplifying systemic insulin resistance [6].

2.3 Neurohormonal integration

The sympathetic nervous system, hypothalamic–pituitary axis, and pancreatic islet hormones synchronise organ responses. Catecholamines increase adipose lipolysis and hepatic glycogenolysis to provision active muscle [3]. Cortisol and glucagon favour gluconeogenesis and proteolysis during stress, whereas insulin and IGF-1 promote anabolism and substrate storage [3]. Myokines feed back into these axes – for example, contraction-induced IL-6 can increase GLP-1 secretion, augmenting insulin release, while apelin improves endothelial function and skeletal muscle perfusion [5, 7].

3. Muscle–liver crosstalk

3.1 Normal physiology

In fed states, skeletal muscle plays a crucial role in removing circulating glucose in response to insulin, helping to normalise blood glucose concentrations [2]. This reduces the liver's workload by decreasing the need for hepatic glucose production [2]. During fasting or prolonged caloric restriction, the liver assumes primary responsibility for maintaining euglycaemia by depleting glycogen stores (glycogenolysis) and initiating gluconeogenesis, utilising substrates such as lactate, glycerol, and amino acids derived from muscle proteolysis [3]. Muscle-derived metabolites such as alanine and lactate serve as essential precursors for hepatic glucose synthesis, facilitating a tightly regulated cycle that supports energy balance [3].

3.2 Exercise and acute responses

Physical activity transiently amplifies this crosstalk. Contracting muscles release cytokines like IL-6, which act hormonally to stimulate hepatic glucose output, thus providing additional energy substrates during increased muscular demand [6]. Simultaneously, lactate produced in active muscles is transported back to the liver via the Cori cycle, a process that prevents excessive accumulation of lactate and spares muscle protein from unnecessary breakdown [9]. This recycling mechanism ensures energy-efficient communication, preserving muscle mass while supporting metabolic flexibility [9].

3.3 Pathophysiological disruptions in chronic disease

In chronic metabolic diseases, such as insulin resistance and T2DM, the architecture of muscle–liver crosstalk often becomes maladaptive. Impaired glucose uptake by skeletal muscle results in persistent hyperinsulinaemia, which stimulates hepatic de novo lipogenesis. This process leads to the accumulation of hepatic fat, culminating in NAFLD [11]. Elevated insulin levels also promote increased lipolysis in adipose tissue, further supplying substrates for hepatic fat synthesis, thereby exacerbating liver steatosis and inflammation [11]. This cycle perpetuates systemic metabolic dysfunction and exacerbates insulin resistance across tissues.

In advanced liver diseases like cirrhosis, the typical role of the liver in glucose and ammonia metabolism becomes compromised. The liver's impaired capacity to detoxify ammonia shifts this responsibility to skeletal muscle, which uses BCAAs – mainly leucine, isoleucine, and valine – to synthesise glutamine. During this process, muscle proteolysis is upregulated to supply BCAAs, leading to significant muscle wasting or sarcopenia [3]. This muscle wasting further diminishes the body's ability to detoxify ammonia efficiently, creating a vicious cycle that worsens both liver and muscle health [3].

At a molecular level, impaired muscle glucose uptake increases circulating glucose and insulin, driving hepatic SREBP-1c–mediated lipogenesis, while chronic inflammation and altered myokine profiles further suppress hepatic fatty acid oxidation [11]. These coordinated changes reflect systemic metabolic reprogramming rather than isolated organ dysfunction [11].

3.4 Impact of surgical stress and critical illness

Major surgery and critical illnesses impose additional stress on the muscle–liver axis. The physiological response to stress involves increased secretion of catabolic hormones, such as cortisol and catecholamines, which stimulate muscle proteolysis to supply amino acids necessary for hepatic synthesis of acute-phase proteins and tissue repair [7]. While initially adaptive, sustained or excessive muscle breakdown delays recovery, hampers immune function, and predisposes patients to infections, wound-healing complications, and prolonged hospital stays [7]. Moreover, the heightened inflammatory milieu further disrupts normal metabolic signalling, contributing to systemic cachexia, declining muscle mass, and worsening hepatic function [7].

3.5 Implications for therapeutic strategies

Understanding the intricate mechanisms of muscle–liver crosstalk opens avenues for targeted therapies aimed at restoring metabolic balance. These include nutritional interventions, pharmacologic agents that modulate cytokine signalling, and exercise programmes designed to improve muscle insulin sensitivity and reduce hepatic fat deposition [3]. In surgical and critically ill populations, strategies such as prehabilitation to enhance muscle mass and tailored metabolic support may mitigate adverse outcomes related to this vital axis [3].

In summary, the muscle–liver crosstalk is fundamental to metabolic health, facilitating a finely tuned balance between fuel supply and demand. Disruptions in this axis contribute to the pathogenesis and progression of numerous metabolic and liver diseases, highlighting the importance of integrated approaches to manage and therapeutically target these interconnected organs [3].

4. Muscle–adipose crosstalk

4.1 Normal physiology and healthy crosstalk

Adipose tissue, traditionally viewed as merely an energy storage depot, is now recognised as a dynamic endocrine organ that plays a crucial role in coordinating whole-body metabolism by interacting closely with skeletal muscle. In a metabolically healthy state, white adipose tissue (WAT) secretes a variety of bioactive molecules called adipokines, which help regulate insulin sensitivity, lipid metabolism, and appetite. For instance, adiponectin, an adipokine with anti-inflammatory and insulin-sensitising properties, enhances skeletal muscle's ability to oxidise fatty acids, thereby improving glucose uptake and overall metabolic flexibility. Leptin, another key adipokine, primarily acts on the hypothalamus to modulate appetite and energy expenditure, balancing caloric intake with energy utilisation [12].

During physical activity, contracting skeletal muscles release several myokines – such as IL-6, β-aminoisobutyric acid (BAIBA), and irisin – that serve as signals to adipose tissue, promoting beneficial adaptations. IL-6, in particular, stimulates lipolysis in WAT, leading to the release of free fatty acids used as fuel by muscles. Additionally, myokines like irisin and BAIBA promote the browning of WAT, transforming beige adipocytes that have higher mitochondrial content and thermogenic capacity akin to brown fat. This browning process increases energy expenditure and contributes to weight regulation and metabolic health [4].

4.2 Disrupted crosstalk in obesity

In obesity, the normal harmonious interaction between muscle and adipose tissue becomes maladaptive. WAT becomes inflamed and fibrotic due to immune cell infiltration and cytokine dysregulation. This inflamed WAT secretes pro-inflammatory cytokines such as tumour necrosis factor-alpha (TNF-α), interleukin-1 beta (IL-1β), and monocyte chemoattractant protein-1 (MCP-1). These inflammatory mediators impair insulin signalling pathways in skeletal muscle, reducing glucose uptake and exacerbating insulin resistance [12]. Furthermore, excess adiposity results in lipid spillover – where fatty acids escape from adipose tissue into circulation – leading to ectopic fat accumulation in muscles (myosteatosis). Lipid deposition within muscle fibres impairs mitochondrial function, causes mitochondrial stress, and reduces insulin sensitivity. This cycle reinforces systemic insulin resistance and metabolic dysfunction, promoting a vicious circle of worsening obesity, muscle lipid infiltration, and metabolic derangement [11]. In parallel, excess lipid delivery to muscle increases mitochondrial reactive oxygen species (ROS) generation and disrupts redox signaling, further impairing insulin action and reinforcing inflammatory feedback loops between adipose tissue and skeletal muscle [11, 12].

4.3 Additional impacts and feedback loops

The compromised muscle–adipose communication in obesity also diminishes the production of beneficial myokines, further reducing lipolytic and browning signals to adipose tissue. This impaired signalling perpetuates inefficient energy expenditure, weight gain, and metabolic inflexibility. Conversely, physical activity or interventions aimed at reducing adipose inflammation can restore healthier crosstalk, promoting improved insulin sensitivity, increased energy expenditure, and metabolic resilience [12, 13].

4.4 Therapeutic implications

Targeting muscle–adipose interactions offers promising strategies for metabolic disease management. Exercise training enhances the secretion of beneficial myokines and promotes WAT browning, leading to increased thermogenesis and weight loss. Pharmacologic agents aiming to reduce adipose inflammation or augment adipokine signalling (e.g., adiponectin mimetics or leptin sensitizers) may restore healthy crosstalk. Nutritional interventions that reduce adipose tissue inflammation can also improve muscle insulin sensitivity and overall metabolic health [12, 13]. Muscle–adipose crosstalk is a vital component of metabolic regulation, coordinating energy storage and expenditure in health and mediating disease progression in obesity-related conditions. Understanding and modulating this interaction can pave the way for novel therapies targeting metabolic syndrome and associated disorders [12, 13].

5. Muscle–pancreas crosstalk

Skeletal muscle is the largest insulin-sensitive organ in the body and is responsible for disposing of approximately 70%–80% of postprandial glucose. Consequently, the insulin responsiveness of skeletal muscle has a profound influence on pancreatic β-cell workload and overall glycaemic homeostasis [14]. In the healthy state, insulin promotes glucose uptake into myocytes via GLUT4 translocation, ensuring efficient storage as glycogen or utilisation through glycolysis and oxidative phosphorylation. When this process is efficient, β-cells require only modest insulin secretion to maintain euglycaemia [14].

5.1 Acute exercise effects

Exercise serves as a powerful modulator of this axis. Even a single bout of physical activity increases insulin sensitivity for up to 48 hours by stimulating AMPK and CaMK pathways that drive GLUT4 translocation independent of insulin [9]. Beyond direct effects on muscle, exercise also triggers systemic endocrine responses. For example, contraction-induced release of IL-6 enhances glucagon-like peptide-1 (GLP-1) secretion from intestinal L-cells. GLP-1 not only potentiates glucose-stimulated insulin secretion but also has trophic effects on β-cell mass and function, illustrating how skeletal muscle activity acutely relieves pancreatic burden [10].

5.2 Chronic dysfunction

In contrast, pathological states such as obesity and type 2 diabetes mellitus (T2DM) disrupt this synergy. Chronic lipid oversupply to muscle leads to intramyocellular lipid accumulation, particularly of diacylglycerol and ceramides, which interfere with insulin signalling at the level of IRS-1 and Akt [11, 13]. Inflammatory cytokines (e.g., TNF-α, IL-1β) secreted from adipose tissue and, to a lesser extent, myocytes, further impair insulin signalling [13]. As skeletal muscle becomes resistant to insulin, higher levels of circulating insulin are required to achieve the same degree of glucose disposal [14]. This increased β-cell demand accelerates β-cell exhaustion and apoptosis, contributing to the progressive decline in insulin secretory capacity characteristic of T2DM [3]. Collectively, these findings position skeletal muscle as an upstream regulator of pancreatic β-cell fate, where muscle insulin resistance and altered myokine signalling accelerate β-cell stress, dedifferentiation, and failure [10].

5.3 Myokine regulation

Emerging research highlights the role of myokines in mediating the dialogue between skeletal muscle and pancreatic islets. IL-15 has been shown to enhance glucose uptake and fatty acid oxidation in muscle, indirectly reducing β-cell strain [6]. Conversely, myostatin, a negative regulator of muscle growth, is associated with reduced insulin sensitivity, impaired mitochondrial biogenesis, and increased β-cell workload [5]. Thus, the balance of myokine secretion from skeletal muscle acts as a determinant of pancreatic functional reserve [10].

5.4 Clinical implications

Lifestyle interventions that increase muscle mass and enhance insulin sensitivity – such as resistance training and aerobic exercise – can substantially reduce the risk of T2DM and even induce remission in early disease [14]. Pharmacological strategies targeting myokine pathways (e.g., anti-myostatin antibodies, IL-15 analogues) are being investigated as adjunctive therapies to reduce β-cell stress [5, 6]. These findings reinforce the concept that the metabolic health of skeletal muscle and pancreas is inseparable, with deterioration in one organ inevitably imposing stress on the other [10].

6. Muscle–gut–brain axis

The muscle–gut–brain axis is an emerging paradigm highlighting the integrated communication among skeletal muscle, the intestinal microbiome, and the central nervous system. This triad operates through metabolic, neural, and endocrine pathways, shaping systemic metabolism, immune responses, and cognitive function [6, 14]. Recent work suggests that microbiome-derived metabolites and muscle-derived myokines converge on shared intracellular pathways – including AMPK, PGC-1α, and inflammatory signalling cascades – highlighting a coordinated metabolic dialogue between microbial ecology, muscle energetics, and neural function [5, 6].

6.1 Microbiome-derived metabolites

The gut microbiota produces a range of metabolites that influence skeletal muscle physiology. Short-chain fatty acids (SCFAs), particularly butyrate, propionate, and acetate, are fermentation products of dietary fibre. These molecules enhance muscle mitochondrial biogenesis through PGC-1α activation, improve insulin sensitivity via AMPK signalling, and exert anti-inflammatory effects through regulatory T-cell expansion [6, 14]. Other microbial metabolites, such as indole derivatives from tryptophan, modulate systemic inflammation and muscle oxidative capacity. Secondary bile acids can signal through FXR and TGR5 receptors, influencing lipid metabolism and muscle energy utilisation [6, 14].

6.2 Exercise–microbiome feedback

Physical activity reciprocally shapes the microbiome by enriching taxa that favour SCFA production, such as *Faecalibacterium prausnitzii* and *Roseburia*. This creates a positive feedback loop: exercise enhances SCFA production, which, in turn, improves muscle energetics, endurance, and recovery [6, 14]. Trained athletes consistently demonstrate greater microbial diversity and functional capacity for SCFA synthesis compared with sedentary individuals. Thus, the gut microbiome can be considered a metabolic amplifier of exercise benefits [6, 14].

6.3 Muscle–brain signalling

Skeletal muscle also communicates directly with the brain through myokines. Irisin, cleaved from FNDC5 during exercise, crosses the blood–brain barrier and stimulates BDNF expression, promoting hippocampal neurogenesis, synaptic plasticity, and cognition [6, 10]. Cathepsin B, another exercise-induced myokine, exerts similar neurotrophic effects [10]. These mediators provide a molecular basis for the well-documented cognitive benefits of regular physical activity, including a reduced risk of neurodegenerative diseases [10].

6.4 Neural regulation of muscle

Conversely, central neural outputs regulate skeletal muscle substrate use and function. Sympathetic nervous system activation increases lipolysis and glycogenolysis, providing substrates for contracting muscle [10]. Vagal signalling modulates gut motility and microbial ecology, indirectly influencing muscle metabolism [6]. This bidirectional interplay underscores the interconnectedness of neural circuits, muscle bioenergetics, and microbial ecology [6, 10].

6.5 Dysbiosis and pathology

Disruption of this axis is evident in multiple disease states. In obesity and metabolic syndrome, reduced microbial diversity and depletion of SCFA-producing species impair muscle insulin sensitivity, while endotoxin (lipopolysaccharide) leakage from the gut triggers systemic inflammation and anabolic resistance [6]. In cirrhosis, altered bile acid metabolism and dysbiosis exacerbate muscle wasting (sarcopenia), contributing to frailty and poor transplant outcomes [3]. In inflammatory bowel disease (IBD), chronic inflammation and microbial imbalance similarly compromise muscle metabolism, reducing exercise tolerance and quality of life [6].

6.6 Clinical implications

Interventions aimed at restoring muscle–gut–brain communication are under active exploration. Prebiotic and probiotic strategies to enhance SCFA production, faecal microbiota transplantation in selected disorders, and exercise programmes tailored to metabolic health all represent promising approaches [6]. On the neurological side, the recognition that exercise-induced myokines promote neurogenesis and cognitive resilience has spurred research into their therapeutic potential for Alzheimer's disease, Parkinson's disease, and depression [10].

6.7 Broader perspective

The muscle–gut–brain axis highlights how skeletal muscle is not simply an effector organ of locomotion but a central player in systemic health. Through crosstalk with both the microbiome and the nervous system, muscle participates in regulating metabolism, immunity, and cognition [6, 10]. Disruption of this axis contributes to frailty, metabolic disease, and cognitive decline – conditions that increasingly dominate modern clinical practice [6, 10].

7. Muscle crosstalk in specific disease states

7.1 Obesity and metabolic syndrome

Obesity is marked by ectopic lipid deposition in skeletal muscle, activation of serine kinases that inhibit insulin signalling, and impaired mitochondrial oxidative capacity. Adipose tissue inflammation amplifies these defects through cytokine spillover. The result is reduced muscular glucose disposal, compensatory hyperinsulinaemia, and hepatic lipogenesis culminating in NAFLD. Restoring healthy crosstalk requires calorie deficit, resistance and aerobic training, and – in selected patients – pharmacotherapy that reduces adiposity and improves insulin sensitivity. Exercise re-establishes an anti-inflammatory myokine profile and enhances muscle fatty-acid oxidation, thereby decreasing lipid spillover to the liver [11, 12].

7.2 Type 2 diabetes mellitus

In T2DM, skeletal muscle insulin resistance is the dominant defect driving hyperglycaemia. Reduced GLUT4 translocation, impaired glycogen synthesis, and accumulation of lipid intermediates (diacylglycerols, ceramides) blunt insulin action. The β-cells initially compensate with hyperinsulinaemia but eventually fail. Therapeutics that target muscle – structured exercise, weight loss, metformin, GLP-1 receptor agonists, SGLT2 inhibitors – improve systemic glycaemia partly by enhancing muscle substrate handling. Pre-operative optimisation of glycaemia and muscle conditioning reduces post-operative complications in people with diabetes [3, 14].

7.3 Cancer cachexia

Cancer cachexia is a systemic wasting syndrome driven by tumour and host-derived factors that rewire muscle–organ crosstalk towards catabolism. Proteolysis-inducing factor (PIF), activin A, and inflammatory cytokines activate ubiquitin–proteasome and autophagy–lysosome pathways in muscle, while lipid-mobilising factors promote adipose lipolysis. Amino acids and fatty acids thereby redirected to the liver and tumour support gluconeogenesis and tumour metabolism [3, 7]. Cachexia reduces tolerance to chemotherapy and increases surgical risk. Multi-modal care – nutrition, resistance training, symptom control, and experimental agents such as myostatin/activin inhibitors – aims to restore anabolism and function [1, 5].

7.4 Critical illness and sepsis

Critical illness elicits a hypercatabolic response wherein muscle provides substrates for immune and hepatic demands. Elevated catecholamines, cortisol, and inflammatory mediators drive insulin resistance and rapid loss of myofibrillar protein. Mitochondrial dysfunction reduces ATP production and exercise capacity long after ICU discharge. Early mobilisation, adequate protein (with attention to renal and hepatic function), avoidance of deep and prolonged sedation, and glycaemic control mitigate ICU-acquired weakness. Importantly, these strategies act through the same crosstalk pathways described above – reducing inflammation, restoring insulin sensitivity, and re-establishing healthy myokine profiles [3, 7, 11].

7.5 Chronic kidney disease

CKD exemplifies maladaptive muscle–kidney crosstalk. Metabolic acidosis stimulates muscle proteolysis; uraemic toxins and inflammation suppress protein synthesis and reduce mitochondrial quality control; anorexia and low physical activity worsen muscle loss. Sarcopenia in CKD correlates with frailty, hospitalisation, and mortality. Interventions include correction of acidosis, individualised exercise prescription (aerobic and resistance), and optimised protein–energy intake. Trials of myostatin/activin pathway inhibitors and anti-inflammatory strategies are ongoing [3,7].

8. Clinical and surgical implications

For surgeons and peri-operative teams, skeletal muscle status is a modifiable risk factor and a therapeutic target. Pre-operative sarcopenia predicts higher rates of infection, delayed wound healing, anastomotic complications, and mortality. Computed tomography–derived indices (e.g., skeletal muscle index at L3) and functional measures (hand-grip strength, gait speed) stratify risk. Practical strategies include:

(1) **Prehabilitation:** Four to eight weeks of combined aerobic and resistance training improve cardiorespiratory fitness, insulin sensitivity, and muscle strength [3].

DOI: http://dx.doi.org/10.5772/intechopen.1014646

(2) **Nutrition:** Ensure protein intake of ~ 1.2–1.5 g/kg per day in most surgical candidates (adjust for renal disease), with leucine-rich sources to stimulate muscle protein synthesis; consider peri-operative carbohydrate loading in appropriate pathways [3].

(3) **Inflammation and glycaemia:** Protocolised analgesia, early mobilisation, and strict but safe glycaemic control reduce catabolic signalling [11].

(4) **Medication review:** Avoid or minimise prolonged corticosteroid use where feasible; consider agents that improve metabolic health (e.g., GLP-1 receptor agonists in obese patients) with multidisciplinary input [3].

In the intensive care unit (ICU) and major trauma, preventing and treating ICU-acquired weakness requires early, progressive mobility; adequate protein-energy delivery; vitamin D repletion when deficient; and attention to haemodynamics and oxygen delivery to preserve mitochondrial function. Importantly, these strategies act through the same crosstalk pathways described above – reducing inflammation, restoring insulin sensitivity, and re-establishing healthy myokine profiles [11, 12].

Importantly, these interventions do not act in isolation but recalibrate shared metabolic and inflammatory networks across multiple organs, underscoring why multimodal strategies are consistently superior to single-target approaches [3, 11].

9. Future directions

The study of skeletal muscle crosstalk with visceral organs is rapidly moving from descriptive physiology towards translational application. Three frontiers, in particular, are poised to reshape both surgical and medical practice:

9.1 Integrated multi-omics for personalised medicine

Traditional biomarkers, such as HbA1c or creatinine, provide only a coarse snapshot of systemic health. Advances in multi-omics – encompassing metabolomics, proteomics, epigenomics, single-cell transcriptomics, and microbiomics – now allow high-resolution mapping of inter-organ communication networks [3, 6, 11]. For example, circulating metabolomic profiles can reveal early disruption in muscle–liver lipid flux, while single-cell RNA sequencing of muscle biopsies identifies inflammatory cell infiltration predictive of insulin resistance [3, 6]. In the near future, patient-specific crosstalk 'signatures' may enable clinicians to stratify surgical risk, predict recovery trajectories, or personalise therapies for conditions such as cachexia or sarcopenia [1, 7, 11]. Importantly, these approaches must be integrated with clinical data, imaging, and functional assessment to create holistic risk prediction models rather than isolated laboratory findings [3].

9.2 Pharmacological 'exercise mimetics'

While exercise remains the most effective inducer of beneficial muscle crosstalk, many patients – such as those with advanced heart failure, severe frailty, or perioperative immobility – are unable to engage in sufficient activity [3, 10, 13].

This has fuelled interest in pharmacological 'exercise mimetics': small molecules designed to activate contraction-associated signalling pathways [5, 8]. AMPK agonists, PGC-1α activators, and irisin analogues represent examples under investigation [5, 6]. These agents could augment mitochondrial biogenesis, enhance fatty acid oxidation, or stimulate neurotrophic pathways even in sedentary states [6, 10]. However, the aim is not to replace physical training, which has mechanical, psychological, and systemic benefits beyond molecular signalling, but rather to complement it in populations where exercise is limited or contraindicated [3, 10]. Careful safety evaluation is essential, as inappropriate activation of these pathways risks unintended consequences such as tumour promotion or cardiac arrhythmia [7].

9.3 Precision nutrition as metabolic therapy

Nutrition represents another frontier for leveraging muscle crosstalk [3, 12, 13]. Increasing evidence suggests that not only the quantity but also the quality and timing of protein intake influence muscle–organ signalling [3]. For instance, leucine-rich diets stimulate mTOR activation in skeletal muscle, enhancing protein synthesis and indirectly reducing pancreatic β-cell stress [3, 14]. Micronutrients such as vitamin D, iron, and omega-3 fatty acids modulate mitochondrial function and inflammatory tone, further shaping muscle performance and systemic health [3, 11]. Precision nutrition involves tailoring these interventions to individual metabolic phenotypes – whether through bedside nitrogen balance studies, wearable metabolic monitors, or integration with omics data [3]. In the surgical setting, perioperative nutrition protocols could be personalised to optimise recovery, while in chronic disease, diet may be adjusted to mitigate sarcopenia, cachexia, or frailty [3, 12].

9.4 Equity and accessibility

Across all these innovations, issues of equity and access remain paramount. Multi-omics profiling and novel pharmacotherapies are resource-intensive and risk widening existing health disparities if confined to academic centres or high-income populations. Precision nutrition interventions must be adapted to cultural and socioeconomic contexts rather than imposed as one-size-fits-all regimens. Ensuring equitable implementation will require cross-disciplinary collaboration among clinicians, data scientists, ethicists, and policy makers.

9.5 Broader outlook

Ultimately, the convergence of omics, exercise biology, and nutrition science holds the potential to redefine muscle not merely as an effector organ but as a therapeutic target and systemic regulator. By harnessing its crosstalk with the pancreas, heart, kidney, gut, brain, and beyond, future medicine may shift towards preserving muscle health as a cornerstone of disease prevention, surgical optimisation, and healthy ageing. Together, these advances reinforce the concept of skeletal muscle as a systems-level therapeutic target, where interventions aimed at muscle biology propagate benefits across the liver, adipose tissue, pancreas, gut, and brain.

10. Conclusion

Skeletal muscle operates as a master regulator of systemic metabolism through its endocrine outputs, substrate handling, and neurohormonal integration with visceral organs. Disease states distort these circuits, converting adaptive stress responses into chronic catabolism and organ dysfunction. For clinicians, this understanding supports practical interventions: prehabilitation, targeted nutrition, inflammation control – and motivates research into therapies that restore the salutary signals of exercise. Elevating skeletal muscle from a 'passive' peri-operative risk marker to an active therapeutic target is a timely and tractable opportunity to improve outcomes.

Acknowledgement

The author acknowledges the use of OpenAI for language polishing of the chapter.

Author details

Zena Alani
Independent Scientist, Auckland, New Zealand

*Address all Correspondence to: zena.alani96@gmail.com

References

[1] Argilés JM, Busquets S, Stemmler B, López-Soriano FJ. Cancer cachexia: Understanding the molecular basis. Nature Reviews Cancer. 2014;**14**(11):754–762. DOI: 10.1038/nrc3829

[2] Leal LG, Lopes MA, Batista Jr ML. Physical-exercise-induced myokines and muscle–adipose tissue crosstalk: A review of current knowledge and the implications for the metabolic syndrome. Frontiers in Physiology. 2018;**9**:1307. DOI: 10.3389/fphys.2018.01307

[3] Pedersen BK. Physical activity and muscle–brain crosstalk. Nature Reviews Endocrinology. 2019;**15**(7):383–392. DOI: 10.1038/s41574-019-0174-x

[4] Wackerhage H, Woods NM. Exercise-induced signal transduction and gene regulation in skeletal muscle. Journal of Sports Science & Medicine. 2002;**1**(4):103–114

[5] Severinsen MCK, Pedersen BK. Muscle–organ crosstalk: The emerging roles of myokines. Endocrine Reviews. 2020;**41**(4):594–609. DOI: 10.1210/endrev/bnaa016

[6] Chakravarthy MV, Booth FW. Eating, exercise, and "thrifty" genotypes: Connecting the dots toward an evolutionary understanding of modern chronic diseases. Journal of Applied Physiology. 2003;**96**(1):3–10. DOI: 10.1152/japplphysiol.00757.2003

[7] Lee JH, Jun HS. Role of myokines in regulating skeletal muscle mass and function. Frontiers in Physiology. 2019;**10**:42. DOI: 10.3389/fphys.2019.00042

[8] Pedersen BK, Saltin B. Exercise as medicine: Evidence for prescribing exercise as therapy in 26 different chronic diseases. Scandinavian Journal of Medicine & Science in Sports. 2015;**25**(Suppl 3):1–72. DOI: 10.1111/sms.12581

[9] Hoffmann C, Weigert C. Skeletal muscle as an endocrine organ: The role of myokines in exercise adaptations. Cold Spring Harbor Perspectives in. Medicine. 2017;**7**(11):a029793. DOI: 10.1101/cshperspect.a029793

[10] Hotamisligil GS. Inflammation, metaflammation and immunometabolic disorders. Nature. 2017;**542**(7640):177–185. DOI: 10.1038/nature21363

[11] DeFronzo RA, Tripathy D. Skeletal muscle insulin resistance is the primary defect in type 2 diabetes. Diabetes Care. 2009;**32**(Suppl 2):S157–S163. DOI: 10.2337/dc09-S302

[12] Akimoto T, Pohnert SC, Li P, Zhang M, Gumbs C, Rosenberg PB, Williams RS, Yan Z. Exercise stimulates PGC1α transcription in skeletal muscle through activation of the p38 MAPK pathway. The Journal of Biological Chemistry. 2005;**280**(20):19587–19593. DOI: 10.1074/jbc.M408862200

[13] Son Y, Choi C, Lee J, Park G, Hong KY, Sun W, Lee MO, Lee YH. Immunometabolic crosstalk in adiposetissue remodeling: Mechanisms and therapeutic perspectives. Journal of Obesity & Metabolic Syndrome. 2025;**34**(4):344–361. DOI: 10.7570/jomes25080

[14] Pendergrast LA, Ashcroft SP, Ehrlich AM, Treebak JT, Krook A,

Dollet L, Zierath JR. Metabolic plasticity and obesity-associated changes in diurnal post-exercise metabolism in mice. Metabolism: Clinical and Experimental. 2024;**155**:155834. DOI: 10.1016/j.metabol.2024.155834

Section 2

Molecular Mechanisms and Therapeutic Strategies in Skeletal Muscle

Chapter 4

Gut Microbiota and Skeletal Muscle Interaction: The Role of Short-Chain Fatty Acids, G-Protein Receptor Signaling and Metabolic Adaptation

Hilal Kalkan

Abstract

The gut-muscle axis integrates microbial metabolism with skeletal-muscle bioenergetics and repair. This chapter synthesizes mechanistic and translational evidence centered on short-chain fatty acids (SCFAs: acetate, propionate, butyrate) generated by the fermentation of dietary fiber. After epithelial transport (MCT1/SMCT1), SCFAs signal to muscle via G-protein-coupled receptors – FFAR2 (GPR43), FFAR3 (GPR41), and GPR109A – and through histone-deacetylase inhibition. These pathways recalibrate mitochondrial function and substrate selection, enhance insulin sensitivity and glucose handling, attenuate low-grade inflammation, and support the maintenance of muscle mass and performance. Converging data from germ-free and antibiotic-depletion models, fecal-transfer experiments, and early human interventions indicate that diet and physical activity remodel the microbiota toward SCFA-producing taxa, with corresponding gains in strength, endurance, and glycogen storage. The chapter evaluates intervention levers, including fiber-rich diets, targeted pro/pre/synbiotics, and exercise. It outlines priorities for clinical translation: standardized SCFA exposure metrics, dose–response and context dependence, and trials powered for functional and body-composition outcomes.

Keywords: gut-muscle Axis, skeletal muscle metabolism, short-chain fatty acids, G-protein-coupled receptors, mitochondrial function, prebiotics and probiotics

1. Introduction

The human gastrointestinal (GI) tract represents one of the largest interfaces between the host and environmental factors in the body [1]. The collection of bacteria, archaea, and eukarya colonizing the GI tract is termed the "gut microbiota." It has coevolved with the host over thousands of years to form an intricate, mutually beneficial relationship [2]. The gut microbiota is a complicated ecosystem that includes over 100 trillion bacterial cells and around 3 million distinct genes, highlighting its complexity and significance for human health, consisting of approximately 10 times more bacterial cells than human cells and over 100 times the genomic content (microbiome) compared to the human genome [3, 4]. The gut microbiota is considered a central "organ" because of its direct and indirect roles in host physiology and pathophysiology [5]. A healthy microbiota is characterized by a diverse array of well-represented taxa, including bacterial species, fungi, viruses, and other microorganisms. In addition to these dominant organisms, there is a plethora of lesser-known, less abundant species that, despite their diminished numbers, play crucial and highly active roles in various metabolic processes. These often-overlooked entities within the microbial ecosystem contribute significantly to overall health, impacting digestion, immune function, and numerous biochemical pathways in ways that researchers continue to explore and elucidate [4, 6].

The impact of the gut microbiome on human metabolic and immune health is increasingly being acknowledged in both scientific and clinical discussions [7]. Therefore, gut microbiota composition has been linked to human health and disease [8]. Recent studies have shown that the gut microbiota influences skeletal muscle in various ways [9]. The gut microbiota can influence the host's muscle mass and function by regulating systemic inflammation and immunity, nutrient absorption, and insulin sensitivity, directly or indirectly, thereby realizing the "gut–muscle axis" [10]. The concept of crosstalk between the gut microbiota and skeletal muscle emerged from several early animal studies that demonstrated increases in various short-chain fatty acid (SCFA) species following endurance exercise training. SCFAs are synthesized by beneficial gut bacteria through the fermentation of nondigestible carbohydrates, with acetate, butyrate, and propionate identified as the primary end products [10–12]. In parallel, microbial metabolism generates secondary bile acids (via deconjugation/dehydroxylation of host bile acids) and tryptophan-derived metabolites (indole-3-propionate, indole-3-acetate, kynurenines). These metabolite classes act as endocrine signals to skeletal muscle via receptors such as TGR5/FXR (bile acids) and AhR/GPR35/PXR (tryptophan), thereby modulating mitochondrial function, insulin sensitivity, substrate use, and inflammation, as depicted in **Figure 1**. With advanced multidisciplinary techniques, such as high-throughput sequencing and bioinformatics [13], more than 9.9 million microbial genes have been identified in human feces [14], including more than 1,000 bacterial species [15]. The alterations in the gut microbiota composition, named "dysbiosis" – a term used to describe alterations in the composition of gut microbiota – have been shown to lead to a reduction in microbial biodiversity and a disruption of the metabolic functions carried out by these microorganisms. This imbalance can have profound implications for the host's overall health, as it affects not only the digestive system but also the pathophysiology of various peripheral organs. Such disturbances in the gut microbiome may contribute to inflammatory responses, metabolic disorders, and even neurodegenerative diseases, highlighting the critical role of gut health in maintaining systemic wellness. Understanding the mechanisms by which dysbiosis alters microbial metabolism and impacts organ function is essential for developing targeted therapeutic strategies to restore a healthy microbiome and prevent related health issues.

To structure the discussion, this chapter starts by conceptually defining the gut–muscle axis, merging experimental observations with new translational insights. It then delves into signaling mechanisms centered around SCFAs, highlighting

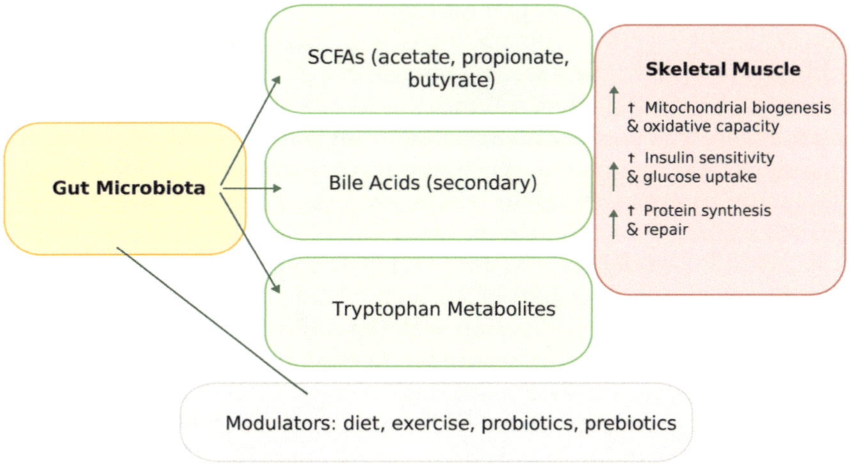

Figure 1.
Gut–muscle axis. Gut microbiome-produced metabolites regulate skeletal muscle. SCFAs (acetate, propionate, butyrate), secondary bile acids, and tryptophan-derived metabolites signal to muscle (green arrows), promoting mitochondrial biogenesis/oxidative capacity, insulin sensitivity and glucose uptake, and protein synthesis/repair, while reducing inflammation and atrophy. Modulators, such as diet, exercise, probiotics, and prebiotics, shape the composition of the microbiota and metabolite output.

receptor-mediated pathways and their subsequent metabolic effects on skeletal muscle. The next sections examine how gut-derived metabolites affect mitochondrial bioenergetics, substrate utilization, inflammatory responses, and exercise adaptations. The concluding section discusses translational perspectives, lingering questions, and priorities for future research.

Instead of adhering to formal systematic-review standards, the literature was carefully selected to encompass both mechanistic depth and translational significance. Particular emphasis was placed on studies utilizing germ-free and antibiotic-depletion models, cellular systems, microbial transplantation techniques, dietary and probiotic approaches, and human exercise scenarios. Research demonstrating mechanistic causality was prioritized, while associative findings were approached with appropriate caution. This method allows for the integration of various experimental systems while maintaining a critical assessment of existing limitations.

2. The potential role of gut microbiota in the gut–muscle axis

Skeletal muscle is one of the largest organs in the human body, accounting for approximately 40–50% of total body weight. It generates heat, controls blood sugar levels, stores amino acids, and modifies the body's physiological traits [16]. So far, a growing number of studies have discovered the relationship between muscle and gut microbiota, both in vitro and in preclinical studies. For example, Fielding et al. compared the gut microbiome of high- and low-functioning (HF, LF) older adults. By transferring fecal samples from two older adults (HF and LF) into germ-free mice (GFM), which lack a gut microbiota, they showed a remarkable increase in grip strength in HF mice compared to LF mice [17]. Notably, a randomized controlled double-blind study showed that prebiotic supplementation, which is a group of nutrients degraded by gut microbiota, increased grip strength in older adults over 65 years old [18]. However, there was no remarkable change in grip strength between the two groups before the intervention.

In contrast, over 13 weeks with prebiotic administration, the experimental group showed a significant increase in grip strength compared with the placebo group. Notably, prebiotics are essential for feeding other beneficial bacteria and promoting their growth and reproduction. On the other hand, probiotics are beneficial bacteria primarily found in our digestive system [19]. The combination of probiotics and prebiotics is called synbiotics [20].

In support of this, supplementation with the prebiotic Faecalibacterium prausnitzii has been shown to increase muscle mass by enhancing mitochondrial respiration, reducing inflammation, and improving insulin sensitivity [21]. In recent decades, the use of probiotics has increased among athletes due to their ability to enhance performance and reduce postexercise fatigue [22]. The study by Lee et al. demonstrated the effects of Lactobacillus salivarius in mice, showing that a 4-week administration of Lactobacillus significantly improved muscle strength and endurance capacity, as well as increased muscle glycogen storage [23]. Moreover, they also conducted a transfer of Bifidobacterium longum subsp. and Bifidobacterium longum OLP-01 (which was extracted from the intestine of the 2008 women's 48 kg weightlifting gold medal champion) into the mice. Over the treatment period of four weeks, glycogen storage, muscle strength, and endurance capacity were all shown to increase significantly [24]. Treatment with antibiotics in the same groups resulted in a reduction of the mice's exercise endurance, which was related to the muscle's ability to utilize glycogen for energy production [25].

3. The role of gut microbiota in mitochondrial and energy metabolism of skeletal muscle

The maintenance of healthy mitochondrial function plays a pivotal role in combating muscle quality and capacity. More importantly, gut microbial metabolites exert numerous regulatory effects on skeletal muscle mitochondrial function [16]. Dysfunctional mitochondria have been linked to muscle atrophy and weakness [26]. The levels of gut microbiota and skeletal muscle mitochondria are connected to the production of insulin-like growth factor 1 (IGF-1) [27]. The study by Lahiri et al. demonstrated that the skeletal muscle of GFM, which lack gut microbiota, compared to the skeletal muscle of pathogen-free mice (PFM), which possess gut microbiota, displays signs of muscle atrophy, decreased expression of IGF1, and reduced mitochondrial function. Notably, the transplantation of the gut-microbiome from PFM into GFM resulted in an increase in skeletal muscle mass, a reduction in muscle atrophy markers, and improved antioxidant capacity in GFM [10]. IGF-1 is one of the best-characterized growth factors that regulate both anabolic and catabolic pathways in skeletal muscle, playing a critical role in maintaining muscle function. Many studies have investigated the role of IGF1 oversecretion in skeletal muscle function and its contribution to the maintenance of muscle quality [28]. Muscle atrophy involves a decline in muscle antioxidant capacity and function, suggesting that the gut microbiome may regulate the normal function of muscle in terms of antioxidant capacity and mitochondrial function [29].

4. The role of gut microbiota in immune and inflammatory functions in skeletal muscle

A reduced microbial community characterizes the imbalance in the microbiome. It alters the intestinal barrier integrity by facilitating the crossing of endotoxins, such as lipopolysaccharides (LPS), and many other microbial products, for example, the tryptophan derivative indoxyl sulfate, into the circulation. Therefore, the imbalance in microbiota composition leads to the triggering of innate immunity by causing low-grade systemic inflammation and, consequently, results in metabolic and muscular disorders [30]. Notably, both tumor necrosis factor α (TNF-α) and interleukin 6 (IL-6) expression are stimulated by LPS and indoxyl sulfate in immune tissues and myoblasts [31, 32]. Recently, the loss of muscle and weakness has been associated with dysbiosis and systemic inflammation. Higher levels of TNF-α and IL-6 have been observed in the elderly and/or patients with inflammatory diseases, and are linked to a reduction in muscle mass and strength [30]. On the other hand, healthy diets (which are described as high in fiber and low in fat) and the use of specific pre- or probiotics have been reported to display many beneficial effects on intestinal, inflammatory, and metabolic parameters [33, 34].

4.1 The gut microbiota and exercise

The discovery of muscle crosstalk with other organs and tissues is essential to understanding how exercise transmits many of its effects on health and performance [4, 10]. Recent studies have established that contracting muscles release proteins and metabolites with endocrine-like properties. Additionally, gut bacteria respond to exercise challenges and have reciprocal roles in fuel selection and muscle function [35]. Therefore, exercise has been demonstrated to display many benefits for the gut microbiota and also for host health [4, 36]. Hereby, exercise is shown to be one of the potent interventions to restore gut symbiosis [37]. These effects of exercise have been associated with increasing the number of beneficial microbial species and enriching microbial diversity, as well as enhancing SCFA synthesis and carbohydrate metabolism [38]. The study by Estaki et al. has shown a positive relationship between fitness and eubiosis (an optimal balance in both population and diversity of gut microflora). Of note, the increased oxygen uptake in the ramp-maximal exercise test has been detected between 18 and 35 years in healthy humans, and this was linked with increased gut microbiome diversity [39]. Gut microbiome diversity has also been associated with the increase in fecal butyrate production and with high levels of key butyrate-producing taxa (such as Clostridiales, Roseburia, Lachnospiraceae, and Erysipelotrichaceae). Moreover, in the study by Estaki et al., similar results have also been reported in mice voluntarily running on a free wheel [40], and in rats voluntarily running while nurtured with a 25% casein-sucrose diet [41].

Additionally, improvements in gut microbiota composition and physical functioning over six weeks with aerobic exercise training resulted in increased SCFA-production capacity and main fecal SCFAs in young, lean adult humans, to a much higher extent than in obese participants [42]. Of note, with these conclusions, the studies are warranting similar types of research also in older adults. As mentioned above, exercise favors the butyrate-producing species such as Faecalibacterium prausnitzii (which is in the Clostridiales order) [38, 40, 43]; therefore, its decrease has been linked with high frailty [44]. There are many benefits of butyrate; for example, the species F. prausnitzii is shown to promote oxygen detoxification and lower the oxygen tension in the lumen. Moreover, in mice that were fed a standard diet (lean) and a high-fat diet (HFD; obese) for 12 weeks, the study revealed that F. prausnitzii butyrate-producing species were detected only in mice that were exercising on a free-running wheel [40].

Additionally, the running distance and time spent on the wheel by mice, as well as the brisk walking of elderly women, have also been associated with an increased abundance of intestinal *Bacteroides* spp. [45, 46]. *Bacteroides* spp. has been inversely related to obesity, which is linked to high-fat and high-carbohydrate diets [47]. In line with this, mice fed low-fat or HFDs have been shown to have a modest inverse relationship between the ΔCt Bacteroidetes: ΔCt Firmicutes ratio [46].

On the other hand, Lactobacillus (Firmicutes phylum), Streptococcus (Firmicutes phylum), and Bifidobacterium (Actinobacteria phylum) produce lactic acid, a major precursor of gut microbiome metabolites of SCFAs [38]. Some species of Lactobacillus or Bifidobacterium have been found to enhance the integrity of the intestinal barrier through the increase in the expression of tight junction proteins, as well as muscle weight and muscle fiber size [48, 49]. More interestingly, the fecal abundances of both the Lactobacillus and/or Bifidobacterium genera have been shown to increase in rats that are fed ad libitum and freely running on a wheel versus the other groups, which are food-restricted and ad libitum-fed groups but without wheel access [50], and also in normal mice or obese rats exercising moderately on a treadmill [35, 51]. Moreover, perturbed gut microbiota in mice treated with an antibiotic (Ab +) is recovered after the natural reseeding of gut microbes. In the Ab+ gut-microbiota-depleted mice, the endurance running capacity was shown to be decreased, and ex vivo skeletal muscle contractile function was impaired [4, 35, 46, 52]. Notably, the natural reseeding of the mice restored the gut microbiota and reversed both skeletal muscle endurance capacity and contractile function [25, 53]. The results have demonstrated that the communication between these two organ systems is bidirectional, with gut microbiota seemingly critical for optimal muscle function, at least in mice [4]. These studies suggest that different types of exercises have varying effects on muscles (**Table 1**). Therefore, exercise is recommended to improve muscle strength, enhance muscle function, and improve muscle structure at the same time [38, 54].

4.2 The influence of diet on gut microbiota and muscle growth/function

The influence of gut microbiota in pathophysiological conditions has been linked to insufficient protein consumption, which is characterized by muscle wasting and is related to gut microbiota immaturity and dysbiosis [55, 56]. The study by Bäckhed et al. revealed that resistance to obesity, induced by a high-fat, high-sugar Western diet for five to eight weeks, is linked to reduced hepatic glycogenesis, increased fatty acid oxidation in the gastrocnemius muscle, and increased locomotor activity. The higher levels of AMP-activated protein kinase (AMPK) and Acetyl-CoA carboxylase (ACC) phosphorylation, as well as carnitine palmitoyl-transferase activity, are associated with higher levels of intestinal production of the lipoprotein lipase (LPL) inhibitor fasting-inducible adipose factor (FIAF) and higher serum triglycerides in the muscle. Therefore, the study highlights the importance of the cooperation of organs in the gut-microbiota-muscle axis [57]. Moreover, apart from diet-induced obesity, a few studies have demonstrated that probiotics also elicit metabolic and

Key points(exercise → microbiome → muscle)	Summary	Ref.
Exercise aligns with eubiosis	Higher cardiorespiratory fitness is associated with greater gut microbial α-diversity and an enrichment of butyrate-producing taxa (Clostridiales, Roseburia, Lachnospiraceae, Erysipelotrichaceae).	[38–40]
Rodent concordance	Voluntary wheel running increases microbial diversity; Faecalibacterium prausnitzii is detected only in exercising mice under normal- and high-fat diets; similar exercise-linked shifts are observed in rats on a casein–sucrose diet.	[40, 41]
Training effects differ by adiposity	Six weeks of aerobic training increase predicted SCFA production capacity and fecal SCFAs, with larger gains in lean than in obese adults.	[42]
Frailty linkage	Lower Faecalibacterium prausnitzii is associated with greater frailty.	[43, 44]
Activity load signal	Abundance of Bacteroides spp. positively correlates with the distance/ time of physical activity and is inversely associated with obesity-related markers.	[45–48]
Lactate → SCFAs pathway	Exercise enriches Lactobacillus and Bifidobacterium (lactate producers); lactate supports cross-feeding to SCFA synthesis and is associated with tighter intestinal junctions.	[49–51]

Table 1.
The impact of exercise on gut microbiota and its effects on muscle function and metabolic health. This table outlines significant findings from both human and animal studies that illustrate how physical activity affects the composition and function of gut microbiota, and how these alterations are connected to skeletal muscle physiology. The evidence features enhanced microbial diversity, increased SCFA-producing taxa, and associations with metabolic adaptation, body fat, and frailty. Both observational and interventional studies are included to emphasize consistent trends across different models.

weight benefits in healthy humans and animals, where they represent a potential strategy to restore muscle function during chronic diseases [58–61]. Hence, the modification of microbiome function might increase the percentage of lean body mass. However, in a few studies, researchers have evaluated probiotics' potential roles in muscle capacities and have found scattered yet encouraging results [55]. Additionally, in murine models, the administration of *L. Plantarum* has shown an increased muscle mass and oxidative muscle phenotype linked to muscle endurance capacity in an exhaustive swimming test and decreased oxidative lesions [55].

4.3 The role of SCFAs in skeletal muscle metabolism and function

The gut microbiota plays a vital role in host health through a range of physiological functions [62, 63]. Beyond these points, SCFAs, including acetate, propionate, and butyrate, can reach the systemic circulation at micromolar concentrations. Therefore, SCFAs can affect the physiology of the host. These subsets of SCFAs include fewer than six carbons, such as acetate (C2), propionate (C3), and butyrate (C4) [64].
A significant body of evidence suggests their well-established effects on whole-body homeostasis [55, 65]. Moreover, the beneficial effects of SCFAs on skeletal muscle function have been described in several studies. SCFAs partly mediate the possible link between gut microbiota and skeletal muscle. The existence of their receptors supports the role of SCFAs in skeletal muscle tissue. In recent studies, multiple roles of SCFAs have been reported and have received further interest **(Table 2)**.

Intriguingly, an important set of molecular targets for SCFAs is defined as G-protein-coupled receptors (GPCRs), which are a large family of

SCFAs	Receptors	Primary roles	Ref.
Acetate (C2)	GPR43, GPR41, OLFR78	Muscle growth and maintenance Mitochondrial function Glucose metabolism Antioxidant and antiinflammatory effects	[107, 108]
Propionate (C3)	GPR43, GPR41, OLFR78	Glucose uptake Oxidative stress Antiinflammatory effects Endurance	[109, 110]
Butyrate (C4)	GPR109A	Mitochondrial biogenesisLipid Metabolism Antiinflammatory effects	[106, 111]

Table 2.
A summary of SCFAs and their receptor-mediated functions in skeletal muscle metabolism and performance. This table outlines the key SCFAs (acetate, propionate, and butyrate), their primary receptors (including GPR41/FFAR3, GPR43/FFAR2, GPR109A, and OLFR78), and their documented roles in skeletal muscle physiology. These roles encompass the regulation of mitochondrial function, glucose metabolism, lipid metabolism, inflammatory responses, and muscle adaptation. Evidence supports receptor-mediated signaling pathways linking microbiota-derived metabolites to systemic metabolic and muscle-related effects.

seven-transmembrane proteins [66, 67]. The GPCRs are activated by internal signal transduction cascades through binding to different ligands. These receptors are identified as a new group of metabolite receptors. The subfamily of GPCRs consists of free fatty acid receptor-2 (FFAR2, also called GPR43) and - 3 (FFAR3, also called GPR41), GPR81, GPR109A, GPR109B, and OLFR78 [68–71]. FFAR2/3 are abundantly expressed on intestinal cells but are also differentially expressed in other cell types such as the intestine [72, 73], adipose tissues [74, 75], pancreas [76], bone marrow [72], liver [77], muscle [78, 79], spleen [80], lungs [81], heart [82], and brain [83] in the host. Both FFAR2/3 receptors activate the heterotrimeric G-coupled protein intracellularly by binding with endogenous SCFAs at the cell surface of specific tissues. Both FFAR2 and FFAR3 are characterized as seven-transmembrane (7TM) spanning proteins, consist of ~ 2% of the whole human and mouse genomes, respectively, and are coupled with Gαi/o-coupled signaling [84]. Only FFAR2 is associated with Gαq/11-coupled signaling [85]. Therefore, based on their expression patterns, structural importance, and activation by gut microbiota metabolites at different tissue levels, these receptors offer a new scope for investigating their role in human diseases.

GPR109A is predominantly expressed in adipocytes of white and brown adipose tissue. It is expressed to a lesser extent in keratinocytes and immune cells, including dermal dendritic cells, monocytes, macrophages, and neutrophils [86]. The striking expression of GPR109A and GPR109B in various immune cells raises the question of what physiological or pathophysiological role these receptors play in leukocytes. GPR81 appears to be expressed principally in adipose tissue [86, 87], although limited expression is also evident in the liver and kidney, as well as in skeletal muscle. Olfr78 is found in vascular resistance beds in various tissues, as well as in the renal afferent arteriole [88].

SCFAs are generated in the colonic lumen at an approximate concentration of 100 mM, exhibiting a composition ratio of 60:20:20 for acetate, propionate, and butyrate, respectively [89]. Concentrations of SCFAs are highest in the colonic lumen and almost negligible in peripheral blood [90], suggesting that they act locally on the epithelium and immune cells present in the colon to induce health-promoting effects.

After being produced in the colonic lumen, SCFAs are transported across the epithelium by diffusion, low-affinity transport mechanisms such as HCO3 –/SCFA exchange (KM 25 mM), medium-affinity transport mechanisms involving monocarboxylate transporter 1 (MCT1, KM 2.6 mM), or via high-affinity transport mediated by sodium-coupled monocarboxylate transporter 1 (SMCT1 or SLC5A8, KM ~ 50 μM) into the colon [91–93]. Upon entering the colonic tissue, a significant portion of SCFAs undergoes metabolic conversion by the colonic epithelial cells. These cells transform SCFAs into lipids or ketone bodies, particularly compounds such as β-hydroxybutyrate and acetoacetate [94, 95], resulting in metabolic activity and a notable reduction in SCFAs that can exit the colonic tissue and subsequently enter the portal circulation [96]. SCFAs reaching the liver are metabolized into lipids. Therefore, negligible amounts of butyrate (4–10 μM), propionate (~4–10 μM), and acetate (~100 μM) are present in peripheral blood, contributing to the body's energy reserves and metabolic functions [97]. EC50 of acetate to induce Ca++ flux through GPR43 is ~ 52 μM. Thus, only acetate is present in a concentration high enough to induce GPR43-dependent signaling in peripheral tissues [98]. Understanding the complex molecular mechanisms underlying the protective effects of SCFAs on colonic health has been challenging because they interact with multiple signaling molecules expressed by colonic epithelial cells and immune cells [89]. The most widely studied targets of butyrate and propionate are histone deacetylases (HDACs). GPR41 and GPR43 share 43% amino acid identity [98] and bind to all three SCFAs: acetate, propionate, and butyrate. While GPR43 prefers SCFAs such as acetate and propionate, the reverse holds for GPR41 [99]. Thus, GPR41 exhibits a notably stronger affinity for butyrate and propionate compared to acetate, indicating its specialized role in recognizing specific SCFAs. In contrast, while GPR109A and GPR109B demonstrate an impressive 96% similarity at the protein level, it is important to note that GPR109B lacks the ability to bind to any SCFAs, setting it apart functionally from its close counterpart [96, 100].

In contrast to GPR41 and GPR43, among SCFAs, only butyrate activates GPR109A [100, 101]. Beta-hydroxybutyrate, produced during starvation as a result of ketogenesis, is also a ligand of GPR109A [100]. The colonic epithelium plays a vital role in maintaining gut health, and it has been demonstrated that this tissue can synthesize β-hydroxybutyrate from butyrate, a SCFA renowned for its beneficial effects on intestinal function [102, 103]. As a result, in the colonic environment, the receptor GPR109A can be activated by both butyrate and its derivative, β-hydroxybutyrate. This raises interesting questions about the mechanisms at play: it would be particularly intriguing to investigate whether some of the positive effects attributed to butyrate's role in promoting colonic health are, in fact, mediated by β-hydroxybutyrate. Exploring this relationship could offer valuable insights into the complex interplay of metabolites in the colon and their implications for overall GI well-being. SCFAs have been detected in the ileum at a concentration of around 10 mM. Nonetheless, it remains unclear whether these substances are generated locally in the ileum or are due to reflux from the colon or cecum. Notably, the concentration of SCFAs present in the ileum is more than sufficient to activate signaling through GPR41 and GPR43 [96, 104, 105]. In addition to supporting evidence, experimental findings from dystrophic models offer direct proof of SCFAs' capacity to improve muscle health. A recent study conducted by our team using the mdx mouse model for Duchenne muscular dystrophy revealed that sodium butyrate supplementation improved locomotor skills, preserved autophagy, and reduced inflammatory markers to levels comparable to the corticosteroid deflazacort. These results were linked to the restoration of gut microbiota composition, elevated circulating butyrate levels, and the activation of SCFA-sensing pathways in skeletal muscle. Importantly, sodium butyrate exhibited similar antiinflammatory and proautophagic properties in human myoblasts derived from Duchenne muscular dystrophy (DMD) patients, highlighting the promise of microbiota-targeted SCFA treatment in addressing muscle disorders [106].

4.3.1 The role of SCFAs in lipid metabolism in skeletal muscle

The effects of SCFAs involve multiple aspects of skeletal muscle, including lipid uptake, oxidation, and storage. In L6 myotubes (rat-derived skeletal muscle cells), the administration of 0.5 mM butyrate has been reported to increase fatty acid oxidation by approximately 30% [112, 113]. Similarly, the same effect on fatty-acid uptake has been observed with the administration of 0.5 mM acetate, approximately 30%, in this same cell line. In the gastrocnemius muscle isolated from C57BL/6 J mice, the ex vivo administration of butyrate (5% [w/w]) for 13 weeks after a HFD also showed a robust increase in fatty acid oxidation [113]. A complementary study by Hong et al. demonstrated that after 10 days of butyrate administration, the levels of triglycerides and cholesterol in skeletal muscle decreased. Significantly, the administration of butyrate (5% [w/w]) for 10 months prevented intramuscular lipid accumulation in C57BL/6 J mice fed a standard diet. The skeletal muscle enzymes involved in lipid metabolism support the changes in this process. For example, LpL, an enzyme that hydrolyzes triglycerides in the circulation and provides free fatty acids to cells, has been shown to increase after SCFA administration in rodents [114]. However, this result is not consistent across all studies [115]. Another important marker for intracellular fatty acid hydrolysis is hormone-sensitive lipase, which has been shown to increase after SCFA administration in mice. Similarly, the ACC enzyme, which is involved in fatty acid biosynthesis, has been shown to decrease after SCFA supplementation in both C57BL/6 J mice and L6 myotubes [116, 117]. However, this change has not been observed after 24 hours of SCFA treatment in C2C12 myotubes (mouse-derived differentiated skeletal muscle cells). In summary, the critical role of SCFAs in skeletal muscle is to increase fatty acid uptake and oxidation, as well as to prevent lipid accumulation [117, 118].

4.3.2 The effects of SCFAs on carbohydrate metabolism in skeletal muscle

The administration of SCFAs increases glucose uptake in skeletal muscle. Both acetate and propionate up-regulate insulin-independent glucose uptake in L6 and C2C12 myotubes, respectively [110, 118]. Besides, coadministration of insulin and propionate increases glucose uptake in both myotubes, an effect partly mediated through GPR4110. The involvement of acetate in glycogen synthesis has also been reported in rats [119, 120]. Moreover, the supplementation of acetate for eight weeks in a standard-fed diet showed a 10-fold increase compared to the control group fed only a standard diet in KK-A(y) mice, a model of obesity and type 2 diabetes[121]. However, glycogen synthase gene expression or enzymatic activity does not appear to be affected by SCFAs [122, 123]. Therefore, SCFAs might regulate glycogen synthesis by inhibiting glycolysis and increasing gluconeogenesis [119, 123].

Consequently, this leads to the accumulation of glucose 6-phosphate. Likely, the increase in glucose uptake and glycogen repletion is facilitated through SCFAs' influence on GLUT4 expression, which is the major glucose transporter protein in skeletal muscle [124]. Complementarily, the increased expression, both at mRNA and protein levels, has been reported in L6 myotubes after constant treatment with 0.5 mM acetate [118]. In line with this result, both Krüppel-like factor 15 and myocyte-specific enhancer factor 2A levels were also increased after acetate supplementation [125]. Somehow, the effects of SCFAs on GLUT4 expression are not exhibited by all SCFAs. For instance, 5 mM butyrate administration for 24 hours in L6 myotubes showed no changes in GLUT4 expression at both gene and protein levels [122].

4.3.3 The effects of SCFAs on skeletal and total body mass

Recently, the effects of SCFAs on skeletal and total body mass have been reported by some authors [113, 126–128]. The acute supplementation of butyrate 5% (w/w) in rodents during a HFD suggests the protective effect of butyrate in skeletal muscle [113]. Alterations in skeletal muscle mass have been shown with a concomitant increase in total body mass in control rodents. However, some other studies revealed no changes in absolute skeletal muscle mass between butyrate-treated versus control mice on a HFD. Another research group [126] also did not detect any alterations, either in skeletal muscle mass or in the skeletal muscle-to-total-body mass percentage, in C57BL/6 J mice fed with a HFD with or without acetate treatment. Although SCFAs might not influence total body mass in energy-exceeding conditions, they can instead reflect metabolic stress/demand or energy-restriction situations. Therefore, this suggests the possible effect of SCFAs' involvement in maintaining skeletal muscle mass (**Table 3**). Followingly, Lahiri et al. reported the effect of SCFAs administration, which was shown to reduce dexamethasone-induced atrophy in vitro in C2C12 myotubes. C2C12 cells, derived from murine skeletal muscle cells, are a well-established model for studying muscle regeneration and differentiation in vitro [129]. Besides, they also demonstrated in gut-microbiota-deficient mice that the administered SCFAs increase skeletal muscle mass [10].

5. Discussion

The emerging concept of a gut microbiota-skeletal muscle axis has significantly advanced our understanding of muscle biology, reshaping the microbiome's role as a crucial systemic regulator of metabolic functions, tissue resilience, and physiological adaptations. Far beyond merely serving as a passive community that aids digestion, the gut microbiota functions like a dynamic, endocrine-like organ. This complex microbial community influences host physiology through a diverse array of metabolites, particularly SCFAs, which play critical roles in this intricate interplay. These potent molecules are produced by microbial fermentation of dietary fibers and subsequently enter the bloodstream, where they reach various peripheral tissues and interact with key nutrient-sensing pathways, primarily via GPCRs. This interaction is essential for regulating various fundamental processes within skeletal muscle, including mitochondrial biogenesis, oxidative capacity, substrate selection for energy production, and modulation of inflammatory responses.

Numerous studies employing GF and antibiotic-treated animal models consistently illustrate that disturbances in the microbiota lead to negative impacts on

Ref.	Subjects/models	Approach	Key findings	Additional notes
[130, 131]	Male ICR mice (6 weeks old)	LP10 supplementation	Enhanced fore-limb grip strength and swimming endurance	LP10 reduced inflammation, improved glucose handling, and increased type I muscle fibers in the gastrocnemius
[132]	Drosophila	Lactobacillus plantarum	Greater protein synthesis and muscle anabolism	mTOR pathway activation and increased muscle protein synthesis
[133]	Male ICR mice (6 weeks old)	NCE supplementation	Improved grip strength and swimming time	Altered gut microbial composition, increased tissue glycogen
[134]	Male C57BL/6 J mice (10 weeks old) Low microbiome-accessible carbohydrate diet	Inulin + microbial transplant	Better endurance	Muscle mass not significantly affected; microbial modulation may require additional interventions
[131, 135]	C2C12 Myoblasts, Myotubes	Lactobacillus curvatus CP2,998	Restored myo-tubular diameter	Protected against dexamethasone-induced atrophy by blocking glucocorticoid receptor activation
[131, 136]	Male ICR mice (6 weeks old)	Kefir supplementation	Increased grip strength, swim endurance, and muscle mass	Modulated gut microbiota, raised glycogen stores
[131, 137]	Male C57BL/6 mice (2 and 10 months old)	Lactobacillus casei LC122 or Bifidobacterium longum BL986	Better muscle strength and performance	Strengthened intestinal barrier and reduced inflammation
[138]	SAMP8 mice (12 and 28 weeks old)	Lactobacillus paracasei PS23	Reduced sarcope-nia indicators	Improved mitochondrial activity, lowered proinflammatory cyto-kine production.
[139]	Male C57BL/6 JNarl mice (6 weeks old, germ-free level)	Eubacterium rectale or Clostridium coccoides colonization	Improved swim endurance	The *E. rectale* microbiota may enhance exercise-related physiological adaptations. *L. plantarum* TWK10 and *C. coccoides* microbiota could improve energy expenditure and body composition when com-bined with exercise.
[140]	CL57BL/6 mice (8 and 12 weeks-old)	Veillonella atypica inoculation	Extended tread-mill running duration	Lactate conversion to propionate for energy
[17, 131]	C57BL/6 J mice Germ-free male and female	Fecal transfer from older high-functioning adults	Greater grip strength	Gut microbiota remodeling and improved intestinal barrier
[141]	C57BL/6 N female mice (7 weeks old)	Faecalibacterium prausnitzii supplementation	Increased muscle mass	Enhanced mitochondrial respiration, reduced inflamma-tion, improved gut integrity
[21]	Overweight Women	Endurance Training (6-weeks period)	Moderate changes in gut microbiota com-position and function.	Increased Akkermansia and decreased Proteobacteria occurred without dietary changes. These microbiota shifts were accompa-nied by reduced android fat mass, improved VO_2max, and lower VAP-1 activity, but had minimal effects on systemic metabolites.

Ref.	Subjects/models	Approach	Key findings	Additional notes
[23]	ICR mice Pathogen-free male (6-week-old)	SA-03	Increased glycogen stores and improved performance	Lowered blood lactate, ammonia, and CK levels
[24]	ICR mice, Male (7 weeks old)	OLP-01	Improved grip strength and endurance	Raised SCFA, liver, and muscle glycogen content
[142]	12-week-old male, specific pathogen-free (SPF), germ-free (GF), and Bacteroides fragilis (BF) gnotobiotic C57BL/6 JNarl	Bacteroides fragilis	Increased muscle mass and endurance	Higher antioxidant enzyme activity
[10]	Germ-free mice	SCFA supplementation	Increased muscle mass and grip strength.	SCFA reduced atrophy-related genes (Atrogin-1, MuRF-1)
[128]	C57Bl/6J mice (6-month-old, young) and (6-month-old, old) female	Butyrate supplementation	Prevented hindlimb muscle atrophy	Increased mitochondrial efficiency and glucose metabolism
[18]	Elderly humans	Prebiotic + fructooligosaccharide	Improved strength, endurance, reduced fatigue	Stimulated beneficial bacterial growth, reduced inflammation
[143]	Triathletes	Lactobacillus plantarum PS128	Increased endurance	Altered microbiota composition, increased SCFA
[144]	Adult subjects over 20 years of age	LP10	Greater muscle mass, reduced fatigue	Enhanced aerobic endurance
[145]	Older adult (>65y) men	High-fiber diet	Higher grip strength	Promoted butyrate production
[45]	Older adult (>65y) women	12-week aerobic training	Stronger trunk muscles	Increased microbiota diversity and SCFA production
[146]	Trained male runners	Probiotic capsules	Better fatigue resistance in heat	Taking a multistrain probiotic for four weeks enhanced running endurance under heated conditions.
[147]	Female swimmers	Probiotic yogurt	Enhanced aerobic capacity	Improved maximum oxygen uptake

Table 3.
Representative studies investigating the influence of gut microbiota modulation on skeletal muscle performance and physiology. This table presents a summary of experimental and clinical research examining the effects of microbiota-targeted interventions, such as probiotics, prebiotics, fecal microbiota transfer, and SCFA supplementation, on skeletal muscle function and metabolic outcomes. The evidence includes in vitro experiments, animal studies, and human trials, indicating improvements in muscle strength, endurance, mitochondrial activity, inflammatory response, and metabolic adaptation. Additionally, the table details the related shifts in microbial populations and the mechanistic pathways that contribute to the associated with muscle health.

muscle mass, mitochondrial efficiency, and exercise tolerance. Notably, these adverse effects can be mitigated by restoring the microbiota or supplementing with SCFAs, suggesting a robust protective role for these metabolites in maintaining muscle health and function.

While there is considerable evidence supporting the idea of a gut–muscle link, significant discrepancies remain in the research. For instance, many studies suggest that SCFA supplementation improves muscle mass and endurance, but other studies report no significant changes in overall skeletal muscle mass when a HFD is involved. Additionally, the effects of SCFAs on GLUT4 expression and lipid metabolism vary, influenced by factors such as species, cell type, dosage, and exposure duration. These discrepancies are likely attributable to differences in experimental design, the host's metabolic state, microbial composition, and methodological approaches. It is important to recognize these differences to avoid overgeneralization and to refine mechanistic hypotheses more effectively.

Several mechanisms have shown consistent evidence, including: (i) the stimulation of AMPK through SCFAs and an improvement in mitochondrial oxidative capacity; (ii) signaling that reduces inflammation via GPCR-dependent pathways; (iii) safeguarding against atrophy caused by dexamethasone in both cellular models and germ-free environments; and (iv) the enhancement of butyrate-producing microbial populations due to exercise. These findings, supported by multiple independent experimental models, lay a solid mechanistic foundation for the gut–muscle axis.

In contrast, several areas remain theoretical or under development. These encompass the specific role of distinct microbial species in determining muscle characteristics in humans, the adequacy of circulating SCFA levels to activate peripheral GPCRs, and the long-term effectiveness of probiotic treatments in influencing muscle mass in clinical populations. Addressing these questions necessitates longitudinal, mechanistically informed human studies that integrate microbiome sequencing with muscle phenotype assessment.

However, despite these compelling preclinical findings, the translational aspect of this research to human physiology remains markedly underdeveloped. Much of the mechanistic understanding is derived from animal studies conducted under rigorously controlled experimental conditions; however, human physiology operates in a far more complex and variable context, shaped by factors such as dietary patterns, genetic predispositions, physical activity levels, and comorbidities. Establishing definitive causal relationships in humans presents considerable challenges; current evidence overwhelmingly consists of associative data, underscoring the field's complexity. Additionally, the composition of an individual's gut microbiota is highly personalized, suggesting that no single microbial or metabolic signature universally correlates with optimal muscle function across diverse human populations. This variability complicates the development of standardized therapeutic interventions and highlights the urgent necessity for personalized treatment strategies tailored to individual microbiome profiles.

Among the plethora of microbial metabolites, SCFAs such as butyrate, acetate, and propionate are potent modulators of muscle health. These compounds not only serve as vital energy substrates that fuel muscular activity but also orchestrate complex transcriptional programs that regulate critical functions, including mitochondrial bioenergetics, lipid oxidation, glucose uptake, and glycogen replenishment. Their significant antiinflammatory properties further protect against catabolic processes that can precipitate muscle wasting. Nonetheless, systemic SCFA levels exhibit substantial variability, shaped by factors such as dietary composition, microbial diversity, and overall gut health. This variability raises pertinent questions about the optimal conditions for maximizing the muscle-protective effects of SCFAs.

Lifestyle interventions are among the most accessible strategies for modulating the gut–muscle axis. Diets enriched with fermentable fibers can promote the proliferation of butyrate-producing microbial species, such as Faecalibacterium prausnitzii, which are associated with beneficial metabolic outcomes. Additionally, both aerobic and resistance training have been shown to independently enhance microbial diversity and increase SCFA production, creating opportunities for synergistic effects on muscle metabolism and overall performance. Preliminary studies suggest that integrating dietary modifications with structured physical activity regimens may enhance muscle function and recovery. Furthermore, the potential of probiotic and prebiotic supplementation as an intervention strategy cannot be overstated; preliminary findings suggest improvements in strength, endurance, and recovery. However, it is essential to recognize that individual responses to these supplements can vary greatly, underscoring the need to identify specific microbial and host factors that influence individual responsiveness.

Future research should shift from simply establishing descriptive associations to conducting longitudinal studies specifically aimed at uncovering causal relationships in human subjects. There is a need for additional studies, particularly controlled clinical trials that examine various gut-modulating therapies, involve larger and more homogeneous groups, and include longer follow-up periods. To ensure practicality, these studies should feature appropriate dosages and durations and be designed as randomized, double-blind, placebo-controlled trials to minimize complications. Moreover, forthcoming research should incorporate comprehensive metrics to document dietary composition, physical activity, and other factors that affect the gut microbiota. It is essential to track medication use and the history of prebiotic or probiotic supplement intake. Analyzing the initial gut microbiota could clarify treatment response differences among individuals, identify "responders" and "nonresponders," and provide valuable insights into the underlying mechanisms of action. Standardizing and employing universal methods to assess skeletal muscle strength, mass, and physical performance would enhance data integration for statistical analysis and interpretation. Using integrative multiomics approaches that combine microbiome sequencing, metabolomics, transcriptomics, and comprehensive muscle phenotyping will be crucial for elucidating the complex mechanistic pathways involved and identifying predictive biomarkers of muscle health. Additionally, it is essential to consider the bidirectional nature of the gut–muscle axis: while gut-derived signals can modulate muscle metabolism and activity, the release of myokines may similarly influence microbial ecology. These interactions may create a feedback loop that promotes metabolic homeostasis and overall health.

Abbreviations

GI	Gastrointestinal
GFM	Germ-free mice
HF	High-functioning
LF	Low-functioning
SCFAs	Short-chain fatty acids
PFM	Pathogen-free mice
IGF-1	Insulin-like growth factor 1

LPS	Lipopolysaccharides
TNF-α	Tumor necrosis factor alpha
IL-6	Interleukin 6
VO_2max	Maximal oxygen consumption
GPCRs	G-protein-coupled receptors
FFAR2 (GPR43)	Free fatty acid receptor 2
FFAR3 (GPR41)	Free fatty acid receptor 3
GPR81	Hydroxycarboxylic acid receptor 1
GPR109A	Hydroxycarboxylic acid receptor 2
GPR109B	Hydroxycarboxylic acid receptor 3
OLFR78	Olfactory receptor 78
MCT1	Monocarboxylate transporter 1
SMCT1 (SLC5A8)	Sodium-coupled monocarboxylate transporter 1
HDAC	Histone deacetylase
CK	Creatine kinase
BUN	Blood urea nitrogen
AMPK	AMP-activated protein kinase
ACC	Acetyl-CoA carboxylase
LPL	Lipoprotein lipase
FIAF	Fasting-induced adipose factor
GLUT4	Glucose transporter type 4
mTOR	Mammalian target of rapamycin
ATP	Adenosine triphosphate
VAP-1	Vascular adhesion protein 1
BF	Bacteroides fragilis
Ab+	Antibiotic-treated
HFD	High-fat diet
C2C12	Mouse myoblast cell line

Author contributions

Conceptualization, H.K.; investigation, H.K.; resources, H.K.; writing–original draft preparation, H.K.; writing–review and editing, H.K.; funding acquisition, H.K. The author has read and agreed to the published version of the manuscript.

Funding

This research was supported by a grant to H.K. A postdoctoral fellowship from IUCPQ supported H.K.

Conflicts of Interest

The author declares no conflicts of interest.

Author details

Hilal Kalkan[1,2,3]

1 Research Center of the Quebec Heart and Lung Institute (IUCPQ), Department of Medicine, Faculty of Medicine, University Laval, Quebec City, Canada

2 Canada Excellence Research Chair on the Microbiome-Endocannabinoidome Axis in Metabolic Health, Québec City, Canada

3 Joint International Research Unit MicroMeNu, between Consiglio Nazionale delle Ricerche, Institute of Biomolecular Chemistry and Université Laval, Pozzuoli, Italy

*Address all correspondence to: hilal.kalkan@criucpq.ulaval.ca

References

[1] Di Vincenzo F, Del Gaudio A, Petito V, Lopetuso LR, Scaldaferri F. Gut microbiota, intestinal permeability, and systemic inflammation: A narrative review. Internal and Emergency Medicine. 2024;**19**:275–293

[2] Bäckhed F, Ley RE, Sonnenburg JL, Peterson DA, Gordon JI. Host-bacterial mutualism in the human intestine. Science. 2005;**307**:1915–1920

[3] Thursby E, Juge N. Introduction to the human gut microbiota. Biochemical Journal. 2017;**474**:1823–1836

[4] Hawley JA. Microbiota and muscle highway — Two way traffic. Nature Reviews Endocrinology. 2020;**16**:71–72

[5] Maciel-Fiuza MF, Muller GC, Campos DMS, do Socorro Silva Costa P, Peruzzo J, Bonamigo RR, Veit T, Vianna FSL. Role of gut microbiota in infectious and inflammatory diseases. Frontiers in Microbiology. 2023;**14**:1098386 DOI: 10.3389/fmicb.2023.1098386

[6] Hul MV, et al. What defines a healthy gut microbiome? Gut. 2024;**73**:1893–1908

[7] Fan Y, Pedersen O. Gut microbiota in human metabolic health and disease. Nature Reviews Microbiology. 2021;**19**:55–71

[8] O'Toole PW, Jeffery IB. Gut microbiota and aging. Science. 2015;**350**:1214–1215

[9] Ticinesi A, et al. Aging Gut Microbiota at the Cross-road between nutrition, physical frailty, and Sarcopenia: Is there a gut-muscle axis? Nutrients. 2017;**9**:1303

[10] Lahiri S, et al. The gut microbiota influences skeletal muscle mass and function in mice. Science Translational Medicine. 2019;**11**:eaan5662

[11] Przewłócka K, Folwarski M, Kaźmierczak-Siedlecka K, Skonieczna-Żydecka K, Kaczor JJ. Gut-muscle axis exists and may affect skeletal muscle adaptation to Training. Nutrients. 2020;**12**:1451

[12] Lustgarten MS. The role of the gut microbiome on skeletal muscle mass and physical function: 2019 update. Frontiers in Physiology. 2019;**10**:1435

[13] Kho ZY, Lal SK. The human gut microbiome-A potential controller of wellness and disease. Frontiers in Microbiology. 2018;**9**:1835

[14] Li J, et al. An integrated catalog of reference genes in the human gut microbiome. Nature Biotechnology. 2014;**32**:834–841

[15] Liguori I, et al. Sarcopenia: Assessment of disease burden and strategies to improve outcomes. Clinical Interventions in Aging. 2018;**13**:913–927

[16] Li G, Jin B, Fan Z. Mechanisms involved in gut microbiota regulation of skeletal muscle. Oxidative Medicine and Cellular Longevity. 2022;**2022**:2151191

[17] Fielding RA, et al. Muscle strength is increased in mice that are colonized with microbiota from high-functioning older adults. Experimental Gerontology. 2019;**127**:110722

[18] Buigues C, et al. Effect of a prebiotic formulation on frailty syndrome: A randomized, double-blind

clinical trial. International Journal of Molecular Sciences. 2016;**17**:932

[19] Davani-Davari D, et al. Prebiotics: Definition, types, sources, mechanisms, and clinical applications. Foods (Basel, Switzerland). 2019;**8**:92

[20] Keulers L, et al. Probiotics, prebiotics, and synbiotics to prevent or combat air pollution consequences: The gut-lung axis. Environmental Pollution. 2022;**302**:119066

[21] Munukka E, et al. Six-week endurance exercise alters gut metagenome that is not reflected in systemic metabolism in over-weight women. Frontiers in Microbiology. 2018;**9**:2323

[22] Chen Z, et al. Probiotics supplementation therapy for pathological neonatal Jaundice: A systematic review and meta-analysis. Frontiers in Pharmacology. 2017;**8**:432

[23] Lee M-C, et al. Lactobacillus salivarius Subspecies salicinius SA-03 is a new probiotic capable of enhancing exercise performance and decreasing fatigue. Microorganisms. 2020;**8**:545

[24] Lee M-C, et al. In Vivo ergogenic properties of the Bifidobacterium longum OLP-01 isolated from a Weightlifting Gold Medalist. Nutrients. 2019;**11**:2003

[25] Nay K, et al. Gut bacteria are critical for optimal muscle function: A potential link with glucose homeostasis. American Journal of Physiology-Endocrinology and Metabolism. 2019;**317**:E158–E171

[26] Hyatt H, Deminice R, Yoshihara T, Powers SK. Mitochondrial dysfunction induces muscle atrophy during prolonged inactivity: A review of the causes and effects. Archives of Biochemistry and Biophysics. 2019;**662**:49–60

[27] Zheng X, Qian Y, Wang L. Causal relationship between gut microbiota and insulin-like growth factor 1: A bidirectional two-sample Mendelian randomization study. Frontiers in Cellular and Infection Microbiology. 2024;**14**:1406132. DOI: 10.3389/fcimb.2024.1406132

[28] Aguirre GA, De Ita JR, de la Garza RG, Castilla-Cortazar I. Insulin-like growth factor-1 deficiency and metabolic syndrome. Journal of Translational Medicine. 2016;**14**(3). DOI: 10.1186/s12967-015-0762-z

[29] Chen S, et al. Gut microbiota in muscular atrophy development, progression, and treatment: New therapeutic targets and opportunities. The Innovation. 2023;**4**:100479

[30] Grosicki GJ, Fielding RA, Lustgarten MS. Gut Microbiota contribute to age-related changes in skeletal muscle size, composition, and function: Biological Basis for a Gut-Muscle Axis. Calcified Tissue International. 2018;**102**:433–442

[31] van der Bruggen T, Nijenhuis S, van Raaij E, Verhoef J, Sweder van Asbeck B. Lipopolysaccharide-induced tumor necrosis factor alpha production by human monocytes Involves the Raf-1/MEK1-MEK2/ERK1-ERK2 pathway. Infection and Immunity. 1999;**67**:3824–3829

[32] Frost RA, Nystrom GJ, Lang CH. Lipopolysaccharide and proinflammatory cytokines stimulate interleukin-6 expression in C2C12

myoblasts: Role of the Jun NH2-terminal kinase. American Journal of Physiology-Regulatory, Integrative and Comparative Physiology. 2003;**285**: R1153–1164

[33] Fu J, Zheng Y, Gao Y, Xu W. Dietary fiber intake and gut microbiota in human health. Microorganisms. 2022;**10**:2507

[34] Victoria Obayomi O, Folakemi Olaniran A, Olugbemiga Owa S. Unveiling the role of functional foods with emphasis on prebiotics and probiotics in human health: A review. Journal of Functional Foods. 2024;**119**:106337

[35] Kim SJ, et al. Dietary fat intake and age modulate the composition of the gut microbiota and colonic inflammation in C57BL/6J mice. BMC Microbiology. 2019;**19**:193

[36] Hawley JA, Forster SC, Giles EM. Exercise, the Gut Microbiome and Gastrointestinal Diseases: Therapeutic impact and molecular mechanisms. Gastroenterology. 2025;**169**:48–62

[37] Varghese S, Rao S, Khattak A, Zamir F, Chaari A. Physical exercise and the gut microbiome: A bidirectional relationship influencing health and performance. Nutrients. 2024;**16**:3663

[38] Mach N, Fuster-Botella D. Endurance exercise and gut microbiota: A review. Journal of Sport and Health Science. 2017;**6**:179–197

[39] Estaki M, et al. Cardiorespiratory fitness as a predictor of intestinal microbial diversity and distinct metagenomic functions. Microbiome. 2016;**4**:42

[40] Campbell SC, et al. The effect of diet and exercise on intestinal integrity and microbial diversity in mice. PloS One. 2016;**11**:e0150502

[41] Matsumoto M, et al. Voluntary running exercise alters microbiota composition and increases n-butyrate concentration in the rat cecum. Bioscience, Biotechnology, and Biochemistry. 2008;**72**:572–576

[42] Allen JM, et al. Exercise alters gut microbiota composition and function in lean and obese humans. Medicine & Science in Sports & Exercise. 2018;**50**:747–757

[43] Bressa C, et al. Differences in gut microbiota profile between women with active lifestyle and sedentary women. PloS One. 2017;**12**:e0171352

[44] Saraswati S, Sitaraman R. Aging and the human gut microbiota-from correlation to causality. Frontiers in Microbiology. 2014;**5**:764

[45] Morita E, et al. Aerobic exercise training with brisk walking increases intestinal bacteroides in healthy elderly women. Nutrients. 2019;**11**:868

[46] Evans CC, et al. Exercise prevents weight gain and alters the gut microbiota in a mouse model of high fat diet-induced obesity. PloS One. 2014;**9**: e92193

[47] Ley RE, Turnbaugh PJ, Klein S, Gordon JI. Microbial ecology: Human gut microbes associated with obesity. Nature. 2006;**444**:1022–1023

[48] Varian BJ, et al. Microbial lysate upregulates host oxytocin. Brain, Behavior, and Immunity. 2017;**61**:36–49

[49] Varian BJ, et al. Beneficial bacteria inhibit cachexia. Oncotarget. 2016;7:11803–11816

[50] Queipo-Ortuño MI, et al. Gut microbiota composition in male rat models under different nutritional status and physical activity and its association with serum leptin and ghrelin levels. PloS One. 2013;**8**:e65465

[51] Petriz BA, et al. Exercise induction of gut microbiota modifications in obese, non-obese and hypertensive rats. BMC Genomics. 2014;**15**:511

[52] Qiu J, Cheng Y, Deng Y, Ren G, Wang J. Composition of gut microbiota involved in alleviation of dexamethasone-induced muscle atrophy by whey protein. NPJ Science of Food. 2023;**7**:58

[53] Guida F, et al. Antibiotic-induced microbiota perturbation causes gut endocannabinoidome changes, hippocampal neuroglial reorganization and depression in mice. Brain, Behavior, and Immunity. 2018;**67**:230–245

[54] Turnbaugh PJ, et al. A core gut microbiome in obese and lean twins. Nature. 2009;**457**:480–484

[55] Gizard F, Fernandez A, De Vadder F. Interactions between gut microbiota and skeletal muscle. Nutrition and Metabolism Insights. 2020;**13**:1178638820980490

[56] Subramanian S, et al. Persistent gut microbiota immaturity in malnourished Bangladeshi children. Nature. 2014;**510**:417–421

[57] Bäckhed F, Manchester JK, Semenkovich CF, Gordon JI. Mechanisms underlying the resistance to diet-induced obesity in germ-free mice. Proceedings of the National Academy of Sciences of the United States of America. 2007;**104**:979–984

[58] Mafe AN, et al. A review on probiotics and dietary bioactives: Insights on metabolic well-being, gut microbiota, and inflammatory responses. Food Chemistry: Advances. 2025;**6**:100919

[59] Mozaffarian D. Dairy Foods, Obesity, and Metabolic Health: The role of the food matrix compared with single nutrients. Advances in Nutrition. 2019;**10**:917S–923S

[60] Cai Y, Liu P, Zhou X, Yuan J, Chen Q. Probiotics therapy show significant improvement in obesity and neurobehavioral disorders symptoms. Frontiers in Cellular and Infection Microbiology. 2023;**13**:1178399

[61] Yadav H, Lee J-H, Lloyd J, Walter P, Rane SG. Beneficial metabolic effects of a probiotic via Butyrate-induced GLP-1 Hormone Secretion. Journal of Biological Chemistry. 2013;**288**:25088–25097

[62] Frontera WR, Ochala J. Skeletal muscle: A brief review of structure and function. Calcified Tissue International. 2015;**96**:183–195

[63] O'Connor G, et al. An open-label pilot single-subject study to monitor the impact of a Food-Based enteral formula on faecal short-chain fatty acid concentrations in children admitted to intensive care with sepsis. Clinical Nutrition Open Science. 2024;**53**:1–10

[64] Macfarlane S, Macfarlane GT. Regulation of short-chain fatty acid production. Proceedings of the Nutrition Society. 2003;**62**:67–72

[65] Frampton J, Murphy KG, Frost G, Chambers ES. Short-chain fatty acids as potential regulators of skeletal muscle metabolism and function. Nature Metabolism. 2020;**2**:840–848

[66] Canfora EE, Jocken JW, Blaak EE. Short-chain fatty acids in control of body weight and insulin sensitivity. Nature Reviews Endocrinology. 2015;**11**:577–591

[67] Hernández-Álvarez D, et al. Long-Term moderate exercise combined with metformin treatment induces an hormetic response that prevents strength and muscle mass Loss in Old Female Wistar rats. Oxidative Medicine and Cellular Longevity. 2019;**2019**:3428543

[68] Kimura I, Ichimura A, Ohue-Kitano R, Igarashi M. Free fatty acid receptors in health and disease. Physiological Reviews. 2020;**100**:171–210

[69] Wanders D, Graff EC, Judd RL. Effects of high fat diet on GPR109A and GPR81 gene expression. Biochemical and Biophysical Research Communications. 2012;**425**:278–283

[70] Ahmed EA. The potential therapeutic role of Beta-Caryophyllene as a chemosensitizer and an inhibitor of angiogenesis in Cancer. Molecules. 2025;**30**:1751

[71] Miyamoto J, et al. Nutritional Signaling via Free Fatty Acid Receptors. International Journal of Molecular Sciences. 2016;**17**:450

[72] Tolhurst G, et al. Short-chain fatty acids stimulate glucagon-like peptide-1 secretion via the G-protein-coupled receptor FFAR2. Diabetes. 2012;**61**:364–371

[73] Samuel BS, et al. Effects of the gut microbiota on host adiposity are modulated by the short-chain fatty-acid binding G protein-coupled receptor, Gpr41. Proceedings of the National Academy of Sciences of the United States of America. 2008;**105**:16767–16772

[74] Jocken JWE, et al. Short-Chain fatty acids differentially affect intracellular Lipolysis in a human white adipocyte Model. Frontiers in Endocrinology. 2018;**8**:372

[75] Iván J, et al. The Short-chain fatty acid propionate inhibits adipogenic differentiation of Human Chorion-Derived Mesenchymal Stem Cells Through the Free Fatty Acid Receptor 2. Stem Cells and Development. 2017;**26**:1724–1733

[76] Veprik A, et al. Acetyl-CoA-carboxylase 1 (ACC1) plays a critical role in glucagon secretion. Communications Biology. 2022;**5**:238

[77] Bindels LB, et al. Gut microbiota-derived propionate reduces cancer cell proliferation in the liver. British Journal of Cancer. 2012;**107**:1337–1344

[78] Pluznick JL. Gut microbiota in renal physiology: Focus on short-chain fatty acids and their receptors. Kidney International. 2016;**90**:1191–1198

[79] Mishra SP, Karunakar P, Taraphder S, Yadav H. Free Fatty Acid Receptors 2 and 3 as microbial metabolite sensors to shape host health: Pharmacophysiological View. Biomedicines. 2020;**8**:154

[80] Bindels LB, et al. Ffar2 expression regulates leukaemic cell growth in vivo.

British Journal of Cancer. 2017;**117**:1336–1340

[81] Halnes I, et al. Soluble fibre meal challenge reduces airway inflammation and Expression of GPR43 and GPR41 in Asthma. Nutrients. 2017;**9**:57

[82] Muralitharan R, et al. Gut Microbiota Metabolites Sensed by Host GPR41/43 Protect Against Hypertension. Circulation Research. 2025;**136**:e20–e33

[83] Lee D-H, Kim M-T, Han J-H. GPR41 and GPR43: From development to metabolic regulation. Biomedicine & Pharmacotherapy. 2024;**175**:116735

[84] Priyadarshini M, et al. An Acetate-Specific GPCR, FFAR2, Regulates Insulin Secretion. Molecular Endocrinology (Baltimore, Md.). 2015;**29**:1055–1066

[85] Bjursell M, et al. Improved glucose control and reduced body fat mass in free fatty acid receptor 2-deficient mice fed a high-fat diet. American Journal of Physiology-Endocrinology and Metabolism. 2011;**300**:E211–220

[86] Ahmed K, Tunaru S, Offermanns S. GPR109A, GPR109B and GPR81, a family of hydroxy-carboxylic acid receptors. Trends in Pharmacological Sciences. 2009;**30**:557–562

[87] Rooney K, Trayhurn P. Lactate and the GPR81 receptor in metabolic regulation: Implications for adipose tissue function and fatty acid utilisation by muscle during exercise. British Journal of Nutrition. 2011;**106**:1310–1316

[88] Pluznick JL, et al. Olfactory receptor responding to gut microbiota-derived signals plays a role in renin secretion and blood pressure regulation. Proceedings of the National Academy of Sciences of the United States of America. 2013;**110**:4410–4415

[89] Ganapathy V, Thangaraju M, Prasad PD, Martin PM, Singh N. Transporters and receptors for short-chain fatty acids as the molecular link between colonic bacteria and the host. Current Opinion in Pharmacology. 2013;**13**:869–874

[90] O'Riordan KJ, et al. Short chain fatty acids: Microbial metabolites for gut-brain axis signalling. Molecular and Cellular Endocrinology. 2022;**546**:111572

[91] Okamura A, Emoto A, Koyabu N, Ohtani H, Sawada Y. Transport and uptake of nateglinide in Caco-2 cells and its inhibitory effect on human monocarboxylate transporter MCT1. British Journal of Pharmacology. 2002;**137**:391–399

[92] Mascolo N, Rajendran VM, Binder HJ. Mechanism of short-chain fatty acid uptake by apical membrane vesicles of rat distal colon. Gastroenterology. 1991;**101**:331–338

[93] Hadjiagapiou C, Schmidt L, Dudeja PK, Layden TJ, Ramaswamy K. Mechanism(s) of butyrate transport in Caco-2 cells: Role of monocarboxylate transporter 1. American Journal of Physiology-Gastrointestinal and Liver Physiology. 2000;**279**:G775–G780

[94] Bergman EN. Energy contributions of volatile fatty acids from the gastrointestinal tract in various species. Physiological Reviews. 1990;**70**:567–590

[95] Zambell KL, Fitch MD, Fleming SE. Acetate and butyrate are the major

substrates for de novo lipogenesis in rat colonic epithelial cells. Journal of Nutrition. 2003;**133**:3509–3515

[96] Sivaprakasam S, Prasad PD, Singh N. Benefits of short-chain fatty acids and their receptors in inflammation and carcinogenesis. Pharmacology & Therapeutics. 2016;**164**:144–151

[97] Cummings JH, Pomare EW, Branch WJ, Naylor CP, Macfarlane GT. Short chain fatty acids in human large intestine, portal, hepatic and venous blood. Gut. 1987;**28**:1221–1227

[98] Brown AJ, et al. The Orphan G protein-coupled receptors GPR41 and GPR43 are activated by propionate and other short chain carboxylic acids. Journal of Biological Chemistry. 2003;**278**:11312–11319

[99] Le Poul E, et al. Functional characterization of human receptors for short chain fatty acids and their role in polymorphonuclear cell activation. Journal of Biological Chemistry. 2003;**278**:25481–25489

[100] Taggart AKP, et al. (D)-beta-Hydroxybutyrate inhibits adipocyte lipolysis via the nicotinic acid receptor PUMA-G. Journal of Biological Chemistry. 2005;**280**:26649–26652

[101] Thangaraju M, et al. GPR109A is a G-protein-coupled receptor for the bacterial fermentation product butyrate and functions as a tumor suppressor in colon. Cancer Research. 2009;**69**:2826–2832

[102] Roediger WE. Utilization of nutrients by isolated epithelial cells of the rat colon. Gastroenterology. 1982;**83**:424–429

[103] Martínez-Ruiz M, Robeson MS, Piccolo BD. Fueling the fire: Colonocyte metabolism and its effect on the colonic epithelia. Critical Reviews in Food Science and Nutrition. 2025;1–20. DOI: 10.1080/10408398.2025.2507701

[104] Bourgot CL, et al. Maternal Short-Chain Fructooligosaccharide Supplementation Influences Intestinal Immune System Maturation in Piglets. PLOS ONE. 2014;**9**:e107508

[105] Le Bourgot C, Ferret-Bernard S, Blat S, Apper E, Le Huërou-Luron I. Short-chain fructooligosaccharide supplementation during gestation and lactation or after weaning differentially impacts pig growth and IgA response to influenza vaccination. Journal of Functional Foods. 2016;**24**:307–315

[106] Kalkan H, et al. Targeting gut dysbiosis against inflammation and impaired autophagy in Duchenne muscular dystrophy. EMBO Molecular Medicine. 2023;**15**:e16225

[107] Liu X, Xu M, Wang H, Zhu L. Role and mechanism of short-chain fatty acids in skeletal muscle homeostasis and exercise performance. Nutrients. 2025;**17**:1463

[108] Kobayashi S, et al. Acetate derived from the intestinal tract has a critical role in maintaining skeletal muscle mass and strength in mice. Physiological Reports. 2024;**12**: e16047

[109] Otten BMJ, Sthijns MMJPE, Troost FJ. A Combination of Acetate, propionate, and butyrate increases glucose uptake in C2C12 Myotubes. Nutrients. 2023;**15**:946

[110] Han J-H, et al. The effects of propionate and valerate on insulin responsiveness for glucose uptake in 3T3-L1 Adipocytes and C2C12 Myotubes via G Protein-Coupled receptor 41. PLOS ONE. 2014;**9**:e95268

[111] Gao Z, et al. Butyrate improves insulin sensitivity and increases energy expenditure in Mice. Diabetes. 2009;**58**:1509–1517

[112] Stilling RM, et al. The neuropharmacology of butyrate: The bread and butter of the microbiota-gut-brain axis? Neurochemistry International. 2016;**99**:110–132

[113] Gao Z, et al. Butyrate improves insulin sensitivity and increases energy expenditure in mice. Diabetes. 2009;**58**:1509–1517

[114] Hong J, et al. Butyrate alleviates high fat diet-induced obesity through activation of adiponectin-mediated pathway and stimulation of mitochondrial function in the skeletal muscle of mice. Oncotarget. 2016;7:56071–56082

[115] Den Besten G, et al. Short-chain fatty acids protect against High-Fat Diet-induced obesity via a PPARγ-Dependent Switch From Lipogenesis to Fat Oxidation. Diabetes. 2015;**64**:2398–2408

[116] Issac PK, et al. Insulin signaling pathway assessment by enhancing antioxidant activity due to morin using in vitro rat skeletal muscle L6 myotubes cells. Molecular Biology Reports. 2021;**48**:5857–5872

[117] Pan JH, et al. Acetic acid enhances endurance capacity of exercise-trained mice by increasing skeletal muscle oxidative properties. Bioscience, Biotechnology, and Biochemistry. 2015;**79**:1535–1541

[118] Maruta H, et al. Activation of AMP-activated protein kinase and stimulation of energy metabolism by acetic acid in L6 Myotube Cells. PloS One. 2016;**11**:e0158055

[119] Fushimi T, Sato Y. Effect of acetic acid feeding on the circadian changes in glycogen and metabolites of glucose and lipid in liver and skeletal muscle of rats. British Journal of Nutrition. 2005;**94**:714–719

[120] Nakao C, et al. Effect of acetate on glycogen replenishment in liver and skeletal muscles after exhaustive swimming in rats. Scandinavian Journal of Medicine & Science in Sports. 2001;**11**:33–37

[121] Li X, et al. Acetic acid activates the AMP-activated protein kinase signaling pathway to regulate lipid metabolism in bovine hepatocytes. PloS One. 2013;**8**: e67880

[122] Chriett S, Zerzaihi O, Vidal H, Pirola L. The histone deacetylase inhibitor sodium butyrate improves insulin signalling in palmitate-induced insulin resistance in L6 rat muscle cells through epigenetically-mediated up-regulation of Irs1. Molecular and Cellular Endocrinology. 2017;**439**:224–232

[123] Fushimi T, et al. Acetic acid feeding enhances glycogen repletion in liver and skeletal muscle of rats. Journal of Nutrition. 2001;**131**:1973–1977

[124] Huang S, Czech MP. The GLUT4 glucose transporter. Cell Metabolism. 2007;**5**:237–252

[125] Yamashita H, et al. Effects of acetate on lipid metabolism in muscles

and adipose tissues of type 2 diabetic Otsuka Long-Evans Tokushima Fatty (OLETF) rats. Bioscience, Biotechnology, and Biochemistry. 2009;**73**:570–576

[126] Kondo T, Kishi M, Fushimi T, Kaga T. Acetic acid upregulates the expression of genes for fatty acid oxidation enzymes in liver to suppress body fat accumulation. Journal of Agricultural and Food Chemistry. 2009;**57**:5982–5986

[127] Henagan TM, et al. Sodium butyrate epigenetically modulates high-fat diet-induced skeletal muscle mitochondrial adaptation, obesity and insulin resistance through nucleosome positioning. British Journal of Pharmacology. 2015;**172**:2782–2798

[128] Walsh ME, et al. The histone deacetylase inhibitor butyrate improves metabolism and reduces muscle atrophy during aging. Aging Cell. 2015;**14**:957–970

[129] Schoneich C, Dremina E, Galeva N, Sharov V. Apoptosis in differentiating C2C12 muscle cells selectively targets Bcl-2-deficient myotubes. Apoptosis: International Journal of Programmed Cell Death. 2014;**19**:42–57

[130] Chen Y-M, et al. Lactobacillus plantarum TWK10 Supplementation Improves Exercise Performance and Increases Muscle Mass in Mice. Nutrients. 2016;**8**:205

[131] Zhao J, Huang Y, Yu X. A narrative review of gut-muscle axis and Sarcopenia: The Potential Role of Gut Microbiota. International Journal of General Medicine. 2021;**14**:1263–1273

[132] Storelli G, et al. Lactobacillus plantarum Promotes Drosophila systemic growth by modulating hormonal signals through TOR-Dependent Nutrient Sensing. Cell Metabolism. 2011;**14**:403–414

[133] Chen Y-M, et al. Supplementation of nano-bubble curcumin extract improves gut microbiota composition and exercise performance in mice. Food & Function. 2020;**11**:3574–3584

[134] Okamoto T, et al. Microbiome potentiates endurance exercise through intestinal acetate production. American Journal of Physiology-Endocrinology and Metabolism. 2019;**316**:E956–E966

[135] Katsuki R, Sakata S, Nakao R, Oishi K, Nakamura Y. Lactobacillus curvatus CP2998 prevents dexamethasone-induced muscle atrophy in C2C12 Myotubes. Journal of Nutritional Science and Vitaminology (Tokyo). 2019;**65**:455–458

[136] Hsu Y-J, et al. Kefir supplementation modifies gut microbiota composition, reduces physical fatigue, and improves Exercise Performance in Mice. Nutrients. 2018;**10**:862

[137] Ni Y, et al. Lactobacillus and Bifidobacterium improves physiological function and cognitive ability in aged Mice by the Regulation of Gut Microbiota. Molecular Nutrition & Food Research. 2019;**63**:e1900603

[138] Chen L-H, et al. Lactobacillus paracasei PS23 decelerated age-related muscle loss by ensuring mitochondrial function in SAMP8 mice. Aging. 2019;**11**:756–770

[139] Huang W-C, Chen Y-H, Chuang H-L, Chiu-C-C, Huang-C-C. Investigation of the Effects of Microbiota on Exercise Physiological

Adaption, Performance, and Energy Utilization Using a Gnotobiotic Animal Model. Frontiers in Microbiology. 2019;**10**:1906

[140] Scheiman J, et al. Meta-omics analysis of elite athletes identifies a performance-enhancing microbe that functions via lactate metabolism. Nature Medicine. 2019;**25**:1104–1109

[141] Munukka E, et al. Faecalibacterium prausnitzii treatment improves hepatic health and reduces adipose tissue inflammation in high-fat fed mice. ISME Journal. 2017;**11**:1667–1679

[142] Hsu YJ, et al. Effect of intestinal microbiota on exercise performance in mice. Journal of Strength and Conditioning Research. 2015;**29**:552–558

[143] Huang W-C, Pan C-H, Wei-C-C, Huang H-Y. Lactobacillus plantarum PS128 Improves Physiological Adaptation and Performance in Triathletes through Gut Microbiota Modulation. Nutrients. 2020;**12**:2315

[144] Huang W-C, et al. Effect of Lactobacillus Plantarum TWK10 on Improving Endurance Performance in Humans. Chinese Journal of Physiology. 2018;**61**:163–170

[145] Barger K, Langsetmo L, Orwoll ES, Lustgarten MS. Investigation of the Diet-Gut-Muscle Axis in the Osteoporotic Fractures in Men Study. Journal of Nutrition, Health & Aging. 2020;**24**:445–452

[146] Shing CM, et al. Effects of probiotics supplementation on gastrointestinal permeability, inflammation and exercise performance in the heat. European Journal of Applied Physiology. 2014;**114**:93–103

[147] Salarkia N, Ghadamli L, Zaeri F, Sabaghian Rad L. Effects of probiotic yogurt on performance, respiratory and digestive systems of young adult female endurance swimmers: A randomized controlled trial. Medical Journal of the Islamic Republic of Iran. 2013;**27**:141–146

Chapter 5

RNA Therapy in Muscle Diseases

Dongchao Lu and Jin Li

Abstract

Skeletal muscle diseases pose a serious threat to human health, yet effective pharmacological treatments remain scarce. RNA-based therapeutics are emerging as a promising avenue: an RNA drug that restores muscle function has already been approved and is in clinical use, highlighting the potential of RNA therapy in this approach. This review surveys the current landscape of RNA therapies for muscle disorders, focusing on the correction of aberrant messenger RNA (mRNA) metabolism, RNA modifications, and the therapeutic manipulation of noncoding RNAs in skeletal muscle diseases. We outline RNA-directed strategies in skeletal muscle, identify the advantages and challenges of RNA therapy in muscle diseases, and discuss future directions for the field.

Keywords: skeletal muscle diseases, muscle atrophy, mRNA, miRNA, lncRNA, circRNA, small interfering RNA, gene editing, delivery

1. Introduction

Muscle diseases are characterized by the loss of muscle mass and function, which are primarily categorized into three major types: hereditary, acquired, and neurogenic. This classification encompasses a wide spectrum of conditions, ranging from progressive atrophy caused by congenital defects to functional impairments triggered by acquired inflammation [1–3]. Hereditary muscular diseases arise from genetic defects, often manifesting in childhood or adolescence with gradually worsening symptoms. Representative examples include Duchenne muscular dystrophy (DMD) and facioscapulohumeral muscular dystrophy (FSHD) [4, 5]. Acquired muscular diseases are induced by postnatal factors without obvious genetic predisposition and can occur at any age. Muscle atrophy (muscle wasting), sarcopenia, thyroid myopathy, and diabetic myopathy are typical instances of this category. Neurogenic muscular diseases essentially result from damage to the nerves that innervate muscles. This nerve impairment leads to the loss of muscle innervation, ultimately causing functional abnormalities. Spinal muscular atrophy (SMA) and amyotrophic lateral sclerosis (ALS) fall into this group [6, 7]. At present, there is a lack of effective drug intervention methods.

RNA therapy is emerging as a new class of drugs with significant potential for the treatment of a variety of diseases, including muscle diseases. After antisense-oligonucleotide (ASO)-directed therapy (Spinraza™) and adeno-associated virus-based gene therapy (Zolgensma) were successfully approved for the treatment of

SMA, RNA therapy is increasingly gaining attention in the field of muscle disease treatment.

This article will review the RNA regulation mechanism and therapy strategies in muscle diseases, introduce the current landscape of RNA-based therapies for the treatment of muscle diseases, and discuss their further development.

2. RNA molecules that regulate muscle diseases

2.1 Dysregulated messenger RNA in muscle diseases

Proper regulation of messenger (mRNA) processing, localization, and translation is essential for skeletal muscle health. Skeletal muscle has high transcriptional flexibility. Precise control of mRNA is important for muscle fiber differentiation, contraction, and regeneration. Abnormal mRNA metabolism, including changes in transcription, alternative splicing, RNA localization, and degradation, is common in many muscle diseases. Disruption of these post-transcriptional mechanisms can cause structural instability, metabolic problems, and loss of muscle mass and function.

Among post-transcriptional mechanisms, abnormal alternative splicing is the most studied. Under normal conditions, splicing regulators like MBNL, CELF, RBFOX, and hnRNP (heterogeneous nuclear ribonucleoprotein) proteins control the switch from fetal to adult isoforms of muscle transcripts. This ensures proper muscle development and function [8–10]. In disease, this developmental splicing is disrupted, and the pattern often reverts to embryonic splicing, which is observed in several muscle disorders [11, 12].

In myotonic dystrophy type 1 (DM1), this dysregulation is pronounced. Expanded CTG repeats in the DMPK gene form nuclear RNA foci. These foci trap MBNL proteins and cause CELF1 upregulation. Thousands of transcripts are mis-spliced [13–15]. For example, abnormal splicing of CLCN1 causes myotonia. Misprocessing of BIN1 and TNNT2 disrupts T-tubules and reduces muscle contraction. Therapeutic strategies, such as ASOs or small molecules that release MBNL, show promise in preclinical studies [16].

FSHD is another example of mRNA dysregulation. The disease is caused by epigenetic derepression of the DUX4 gene in the D4Z4 repeat on chromosome 4. Normally silenced, DUX4 is abnormally expressed in patient muscle. Its mRNA encodes a transcription factor that activates early embryonic and apoptotic genes. Even short bursts of DUX4 mRNA can cause oxidative stress, block differentiation, and kill muscle cells [17].

In DMD, mutations in the DMD gene often create premature stop codons. This triggers nonsense-mediated decay (NMD) and prevents dystrophin production. The lack of dystrophin destabilizes the sarcolemma and promotes inflammation, fibrosis, and poor regeneration. Spatial transcriptomics shows regional upregulation of genes like Col1a1, Tgfb1, and Spp1 in dystrophic muscle [18]. ASO–mediated exon skipping can restore the dystrophin reading frame, turning severe DMD into a milder Becker-like phenotype [19].

Dysregulated alternative polyadenylation (APA) represents another layer of mRNA misregulation in muscle disease. Changes in 3′-UTR (untranslated region) length affect mRNA stability, localization, and micro-RNA (miRNA)-mediated repression. MBNL proteins, besides their role in splicing, also regulate APA; their sequestration in DM1 leads to widespread use of alternative poly(A) sites, contributing to aberrant isoform expression and impaired muscle maintenance [16]. Disruption of mRNA transport along cytoskeletal tracks further impairs localized translation at the sarcolemma and neuromuscular junctions. Hildebrandt et al. [20] showed that MBNL proteins can travel with kinesin motors to deliver specific mRNAs to membrane domains, and disease-associated MBNL dysfunction mislocalizes these transcripts, compromising membrane repair and excitation–contraction coupling.

RNA-binding proteins (RBPs) are key regulators of muscle pathology. Mutations or altered expression of TDP-43, FUS, and hnRNPA2B1 disrupt RNA splicing, transport, and stress granule dynamics in inclusion body myositis and distal myopathies. Loss of RBM20 affects TTN mRNA processing, causing dilated cardiomyopathy [21]. In nemaline myopathy, ACTA1 and NEB mutations lead to aberrant splicing and nonfunctional mRNAs [22]. Inflammatory myopathies, including polymyositis and dermatomyositis, show cytokine-driven mRNA dysregulation, with upregulation of interferon-stimulated and UPR-related transcripts [23].

Mitochondrial and metabolic myopathies are also characterized by abnormalities in mRNA processing. The formation of oxidative phosphorylation complexes depends on the coordinated expression of both nuclear- and mitochondrial-encoded mRNAs. Mutations affect the maturation of mitochondrial transcripts such as MT–ND1 and MT–CO3, or in nuclear genes that regulate mitochondrial mRNA stability, like LRPPRC and GFM1. These unexpected gene mutations can impair translation and trigger energy deficiency in muscle fibers [24–26]. These findings emphasize that muscle dysfunction can result not only from structural gene mutations but also from defects in mRNA processing and translation fidelity.

In conclusion, abnormal mRNA metabolism is a key cause of many skeletal muscle disorders. Toxic RNA, splicing errors, mRNA decay, and mislocalization, all disrupt post-transcriptional regulation. This affects muscle differentiation, contraction, and regeneration. Advances in single-cell transcriptomics, long-read sequencing, and RNA therapies are improving our understanding of mRNA-related pathology. They also offer ways to develop targeted treatments for currently incurable muscle diseases. In addition, many RNA-targeted therapies have been explored and applied in the clinic. Antisense oligonucleotides (ASOs) can silence toxic repeat RNAs or fix splicing defects. They are being tested in DM1 and DMD [27]. Small molecules like branaplam and risdiplam, first developed for SMA, can also modulate splicing [28]. Other strategies include CRISPR editing of splicing sites, stabilizing mRNAs with modified nucleotides, and delivering synthetic mRNAs that encode micro-dystrophin or mini-genes. Future studies may focus on restoring RBP networks and proper mRNA transport in multinucleated muscle fibers.

2.2 RNA modification in muscle diseases

RNA modifications refer to a diverse array of chemical alterations occurring on RNA molecules. They constitute a crucial layer of post-transcriptional regulation, a field known as epitranscriptomics. These modifications dynamically modulate RNA stability, localization, splicing, and translational efficiency, thereby governing essential cellular processes [29–31]. In muscle biology, the precise regulation of RNA function is critical for maintaining the integrity of muscle fibers, their contractile capacity, and regenerative potential. Accumulating evidence indicates that dysregulated RNA modification programs are associated with the pathogenesis of various muscular diseases, highlighting their potential as therapeutic targets in future research [32–34].

2.2.1 M^6A (n^6-methyladenosine)

As the most abundant internal modification in eukaryotic mRNA, m^6A is present at an average of three to five sites per mRNA molecule, accounting for 0.1%–0.4% of the total adenosines in mRNA.

N6-methyladenosine (m^6A), a prevalent post-transcriptional modification of mRNA, exhibits distinct distribution preferences: it is enriched in the conserved sequence motif RRACH (where R = A/G, H = A/C/U) and is primarily localized near stop codons, within long internal exons, and in the 5′-UTR under specific conditions [35, 36]. The dynamic regulation of m^6A is mediated by three key classes of regulatory enzymes: "writers," "erasers," and "readers." The writer complex is centered on METTL3 (catalytic subunit) and METTL14 (structural scaffold), which mediate adenosine methylation with the assistance of cofactors such as WTAP [37, 38]. Among the erasers, FTO removes methyl groups through an oxidative mechanism, while ALKBH5 directly catalyzes demethylation. The readers are dominated by the YTH family: YTHDF1 promotes mRNA translation, YTHDF2 accelerates mRNA degradation, and YTHDC1 regulates splicing and nuclear export. Additionally, IGF2BP1/2/3 enhance mRNA stability, thereby exerting functional antagonism against YTHDF2 [39].

In induced pluripotent stem cell-derived motor neurons (iPSC-MNs) from ALS patients, m^6A levels are reduced [40]. By regulating the expression of MYH7, m^6A modification influences muscle fiber types, which may contribute to the pathogenesis of such diseases. For instance, abnormal m^6A methylation can trigger the aberrant conversion of slow-twitch muscle fibers to fast-twitch ones, impairing muscle endurance and strength, and subsequently leading to muscle dysfunction.

Abnormal muscle development or reduced regenerative capacity is closely linked to skeletal muscle disorders, such as muscle atrophy and impaired post-injury repair [41]. By modulating the expression of key genes involved in myogenic differentiation, m^6A modification may affect normal muscle development and post-injury repair processes. For example, FTO regulates the m^6A modification of the GADD45B gene, influencing its mRNA stability and thereby modulating myogenic differentiation-related signaling pathways (e.g., the P38 MAPK pathway) to promote myocyte differentiation and myotube formation [42].

In myoblasts from patients with FSHD, the expression levels of METTL3 and global m^6A levels are elevated. METTL3 promotes the expression of NPNT and maintains myotube homeostasis in a manner dependent on its m^6A catalytic activity. This finding provides a potential molecular target and intervention strategy for delaying skeletal muscle aging or treating age-related skeletal muscle degenerative diseases, such as sarcopenia [43].

Muscle-specific knockout of YTHDF2 impairs muscle growth. YTHDF2 controls the mRNA stability of the ubiquitin ligase ASB2 through SMAD3, thereby regulating the activation of growth-inhibiting gene programs [44].

Other RNA modifications (N^1-methyladenosine (m^1A), 5-methylcytosine (m^5C), N^7-methylguanosine (m^7G), N^4-acetylcytosine (ac^4C), pseudouridine (Ψ), uridylation, adenosine-to-inosine (A–to–I) RNA editing, and modifications of U34 on tRNA wobble) currently lack corresponding research in skeletal muscle diseases.

2.3 Functions of noncoding RNA in muscle diseases

Accumulating evidence shows that noncoding RNAs (ncRNAs) play an important role in regulating distinct steps of muscle atrophy. The first discovery of ncRNA dates back to the 1950s, when George Palade identified ribosomal RNA (rRNA) [45] and Paul Zamecnik isolated transfer RNA (tRNA) [46]. Next, the small nuclear RNAs (snRNAs) and small nucleolar RNAs (snoRNAs) were further observed and identified as modulators in RNA splicing, modification, and other processing events. In addition, these small ncRNAs are now recognized as "housekeeping noncoding RNAs" essential for maintaining basic cellular functions. In the 1970s, researchers first identified circular transcripts in the cytoplasm; however, their extremely low abundance led to their widespread dismissal as mere by-products of RNA processing. During the same period, transcripts originating from the complementary strand of protein-coding genes were detected in bacteria [47]. Because these RNAs bore little resemblance to any known functional RNA, they were simply labeled as "cellular garbage" and largely ignored. Yet, a few forward-thinking pioneers cautioned that "junk does not mean useless" [48]. Since the late twentieth century, the rapid development of molecular technologies – such as high-throughput sequencing and gene editing – has propelled ncRNA research into an unprecedented boom. Beyond the well-known long ncRNAs (lncRNAs), miRNAs, and circular RNAs (circRNAs), a growing array of novel ncRNAs, including enhancer RNAs (eRNAs), tRNA-derived small RNAs (tsRNAs), and PIWI-interacting RNAs (piRNAs), have been successively identified and characterized in recent years. Expanded evidence has proved that ncRNAs govern a spectrum of biological processes – including cell growth, division, and apoptosis – and are indispensable for maintaining normal cellular physiology. In this chapter, we focus on the best-characterized ncRNA species, such as miRNAs, lncRNAs, and circRNAs, and summarize current knowledge of their actions and molecular mechanisms in skeletal muscle atrophy.

2.3.1 MiRNAs in skeletal muscle atrophy

The miRNAs have about 22 nucleotides and are synthesized via a multistep process in the nucleus and cytoplasm. They primarily regulate gene expression at the post-transcriptional level by binding to target transcripts, leading to their degradation or translational suppression [49]. Recent studies have demonstrated that miRNAs act as precise "molecular knobs," governing protein metabolism, mitochondrial content, autophagy, and cell death in skeletal muscle, critically determining both the onset and potential therapeutic reversal of muscle atrophy.

In various atrophy models, such as immobilization, cancer cachexia, and denervation, the expression of miR-29b is generally upregulated. It targets and inhibits IGF-1 and PI3K (p85α), blocks the AKT signaling pathway, activates FoxO3A, induces the expression of Atrogin-1 and MuRF1, and thereby accelerates the degradation of myofibrillar proteins. Based on these findings, local injection of CRISPR/Cas9 to knock out the pre-miR-29b processing site in the gastrocnemius or tibialis anterior (TA) muscle can restore the activity of the AKT–FOXO3A–mTOR axis, significantly alleviate atrophy induced by AngII, immobilization, or denervation, block myocyte apoptosis triggered by AngII, and remarkably improve exercise endurance [50]. In addition, miR-486 activates the PI3K/AKT signaling pathway by inhibiting phosphatase and tensin homolog (PTEN). By increasing protein synthesis and reducing degradation, miR-486 indirectly inhibits the Hippo signaling pathway, thereby promoting cell growth [51].

In denervation-induced muscle atrophy, there is a significant decrease in the number of mitochondria and an increase in autophagosomes. The miR-142a-5p/mitofusin 1 (MFN1) axis is significantly activated, leading to mitochondrial dysfunction, mitophagy, and cell death. Activation of the miR-142a-5p/MFN1 axis disrupts the tubular mitochondrial network and induces excessive mitophagy and cell death; antagonizing this axis can restore the balance of mitochondrial fusion and fission, thereby alleviating the development of muscle atrophy [52]. On the contrary, fasting, nutrient deficiency, or muscular dystrophy reduces the level of miR-378, disrupting the autophagy-apoptosis homeostasis. Overexpression of miR-378 initiates autophagy through the mTOR–ULK1 pathway, targets PDK1 to enhance FoxO-mediated autophagy gene transcription, and directly inhibits caspase-9 to block endogenous apoptosis, thus reducing the susceptibility of myocytes to apoptosis [53]. Muscle-specific miRNAs, including miR-1, miR-23a, miR-133, miR-206, miR-208b, and miR-499, are upregulated in starvation-induced C2C12 myotubes and in mouse models of muscle atrophy. After 45 days of head-down tilt bed rest, serum levels of miR-23a, miR-206, and miR-499 were increased and correlated with the degree of soleus muscle volume loss, suggesting their potential as noninvasive early biomarkers for unloading-induced atrophy [54]. In summary, miRNAs finely regulate skeletal muscle atrophy through multiple targets and pathways, including protein degradation, mitochondrial quality control, and autophagy-apoptosis balance. Targeting miRNAs holds promise for developing RNA-based precision medicine strategies, offering new insights for the prevention and treatment of sarcopenia, cachexia, and muscle atrophy induced by microgravity in spaceflight. The roles of individual miRNAs are summarized in **Table 1**.

2.3.2 LncRNAs in skeletal muscle atrophy

The lncRNAs have become core nodes in the regulatory network of skeletal muscle atrophy due to their functions, such as transcriptional regulation, molecular sponging, and protein interaction. Existing studies have shown that thousands of lncRNAs are abnormally expressed in the state of muscle atrophy, and these molecules regulate muscle homeostasis through multiple pathways [66, 67].

2.3.2.1 Transcription and epigenetic regulation

Studies have demonstrated that lncRNA-MEG3 functions as an "anti-atrophy regulator" in skeletal muscle. Mice with MEG3 knockout exhibit significant muscle atrophy, increased fat infiltration, and reduced mitochondrial ATP production. In contrast, MEG3 overexpression promotes the conversion of muscle fibers to the oxidative type, thereby enhancing muscle mass and strength. The core mechanism underlying this phenomenon is that MEG3 can directly bind to the SUZ12 protein, promoting its liquid–liquid phase separation (LLPS). This process further maintains

Micro-RNA	Target	Pathway	Function	Refs
miR-29b	MuRF-1, MAFbx-1, FoxO3a, P70S6K	AKT, FOXO3A, mTOR, IGF1	CRISPR/Cas9-mediated miR-29b editing could prevent different types of muscle atrophy in mice	[50]
miR-378	Phosphoinositide-dependent protein kinase 1 (PDK1), caspase-caspase 9	mTOR, ULK1	Micro-RNA-378 promotes autophagy and inhibits apoptosis in skeletal muscle	[53]
miR-142a-5p	MFN1	_	Denervation drives skeletal muscle atrophy and induces mitochondrial dysfunction, mitophagy, and apoptosis via the miR-142a-5p/MFN1 axis	[52]
MiR-486	PTEN	PI3K/AKT, HIPPO	Micro-RNA-486 is involved in the development of muscle atrophy caused by aging, immobility, prolonged exposure to microgravity, or muscular and neuromuscular disorders	[51]
miR-146a-5p	IGF-1 R	PI3K/AKT/mTOR	Adipose-derived exosomes ameliorate skeletal muscle atrophy via miR-146a-5p/IGF-1 R signaling	[55]
miR-132-3p	FoxO3	_	Human umbilical cord mesenchymal stem cell-derived exosomes ameliorate muscle atrophy via the miR-132-3p/FoxO3 axis	[56]
miR-486-5p	Foxo1	_	BMSC-derived exosomes inhibit dexamethasone-induced muscle atrophy via the miR-486-5p/Foxo1 axis.	[57]
miR-497-5p	Insr, Igf1r, Pik3r1, Mapk8ip2, CcnD2/CcnE1, Hmga1	Insulin/IGF-1, JAK-STAT	Decreased miR-497-5p suppresses IL-6-induced atrophy in muscle cells.	[58]
miR-628	IRS1	IRS1/Akt/FoxO3a	miR-628 promotes burn-induced skeletal muscle atrophy via targeting IRS1	[59]
miR-29b	IGF1, YY1, Bax, Bcl-2, Caspase-3, FoxO3a, P70S6K, EIF-4EBP1, p62, LC3, UBC	PPARγ, AKT, mTOR, PI3K	The PPARγ/miR-29b axis could induce angiotensin II–induced muscle atrophy	[60]
miR-23a	Rcan-1, MuRF-1, MAFbx	AKT	Delphinidin, by increasing miR-23a expression via upregulation of NFATc3 expression, could correlate with muscle atrophy	[61]
miR-1, miR-23a, miR-133, miR-206, miR-208b, and miR-499	_	_	Serum miR-23a, miR-206, and miR-499 levels may serve as potential biomarkers for the diagnosis of skeletal muscle atrophy	[54]

Micro-RNA	Target	Pathway	Function	Refs
miR-322	MyHC, IGF1R, INSR, GR	_	miR-322 can exacerbate dexamethasone-induced muscle atrophy by targeting INSR and IGF1R	[62]
miR-133a	NF90, NF45, Dmn2, Aif1l, Atp6ap2, Col8a1, Srgap3	_	miR-133a is a myogenic miRNA that promotes muscle maturation. It can target and inhibit the expression of dynamin 2, and its reduced levels induce centronuclear myopathy	[63]
miR-18a	MuRF-1, CTSL, FoxO1, FoxO3a, Myosin, Myogenin	IGF1/PI3K/Akt	miR-18a, via the IGF1/PI3K/Akt axis, could lead to myotube atrophy	[64]
miR-206	HDAC4	_	miR-206 participates in neuromuscular junction regeneration and muscle atrophy	[65]

Table 1.
Effects of miRNAs on the development of skeletal muscle atrophy.

the histone H3 lysine 27 trimethylation (H3K27me3) modification of muscle-specific genes (e.g., Fhl3, Rnf128), ensuring the normal activation of pathways related to muscle development and function [68]. This finding reveals a novel mechanism by which lncRNA mediates epigenetic regulation through modulating protein phase separation, providing a potential target for the treatment of muscle atrophy.

2.3.2.2 Regulation of skeletal muscle development via chromatin interaction

MYH1G-AS is a skeletal muscle-specific lncRNA associated with open chromatin, and there are differentially expressed ATAC-seq peaks in its promoter region. Studies have shown that MYH1G-AS exerts multiple regulatory effects on skeletal muscle development, including promoting myoblast proliferation, inhibiting myoblast differentiation, driving the transformation of muscle fibers from slow-twitch to fast-twitch, and inducing muscle atrophy [69]. Specifically, MYH1G-AS inhibits the stability of FGF18 protein, reduces the interaction between FGF18 and the transcription factor SMARCA5, thereby suppressing the chromatin accessibility of the SMAD4 promoter, and ultimately activates the SMAD4-dependent pathway to regulate the development of chicken skeletal muscle.

2.3.2.3 Involvement in spatiotemporal regulation of muscle regeneration and development

Studies have demonstrated that lncRNAs are crucial for the spatiotemporal regulation of muscle development and regeneration. Functional experiments confirmed that lncMREF is predominantly expressed in differentiated muscle satellite cells and can effectively promote myogenic differentiation and muscle regeneration [70]. At the mechanistic level, lncMREF interacts with the Smarca5 protein to increase chromatin accessibility during the activation and differentiation initiation of muscle satellite cells. This interaction further facilitates the genomic binding of the p300/CBP complex and histone modification H3K27ac, ultimately upregulating the expression of myogenic regulatory factors such as MyoD and driving cell differentiation. These results reveal a novel time-specific epigenetic regulatory mechanism in muscle regeneration and clarify the decisive role of lncMREF/Smarca5-mediated epigenetic programming in muscle cell differentiation.

2.3.2.4 Targeting atrophy-related pathways

Studies have demonstrated that various lncRNAs play crucial regulatory roles in the progression of skeletal muscle atrophy, and some of them hold promise as potential therapeutic targets.

LncRNA-Myoparr acts as a promoter of denervation-induced muscle atrophy. Its expression is significantly upregulated after denervation injury, and it exacerbates atrophy by regulating the expression of myogenin and growth differentiation factor 5 (Gdf5). Moreover, knockdown of Myoparr activates the bone morphogenetic protein (BMP) signaling pathway in denervated muscles, as evidenced by the increased phosphorylation levels of Smad1/5/8. Additionally, Myoparr can modulate the expression of differential genes related to myogenic differentiation and muscle atrophy, indicating that certain lncRNAs can integrate environmental signals with gene expression networks to directly regulate the activation of atrophy markers [71].

Another study showed that overexpression of lncRNA-H19 increases the mRNA expression of atrophy gene 1 (Atrogin-1), and this effect can be reversed by inhibiting miRNA-675 (miR-675). Further mechanistic studies revealed that overexpression of miR-675 aggravates high stress (HS) and starvation-induced muscle atrophy by inhibiting the insulin-like growth factor 1 receptor/protein kinase B (IGF1R/Akt) signaling pathway and promoting the expression of forkhead box O/ atrophy gene 1 (FoxO/Atrogin-1). Conversely, inhibition of miR-675 alleviates atrophy. This suggests that downregulation of H19/miR-675 expression can ameliorate muscle atrophy by regulating the IGF1R/Akt/FOXO signaling pathway [72].

In addition, lncRNA XLOC_015548 exerts an anti-atrophy effect. The lncRNA XLOC_015548 can promote myogenic differentiation and prevent muscle atrophy by regulating Gadd45g expression, activating the MEK/ERK signaling pathway, and inhibiting oxidative stress. Specifically, its overexpression can significantly downregulate the expression of Gadd45g ($p < 0.05$) and promote the translocation of Gadd45g to the cytoplasm, thereby significantly activating the MEK/ERK signaling pathway ($p < 0.05$) to enhance cell proliferation and myotube formation. Meanwhile, it can reduce the level of reactive oxygen species (ROS) ($p < 0.01$) and stabilize mitochondrial membrane potential, thereby alleviating dexamethasone (DEX)-induced oxidative stress. Moreover, the MEK/ERK pathway inhibitor U0126 can partially reverse the above protective effects, confirming the mediating role of this pathway [73].

In summary, lncRNAs with regulatory activity on muscle atrophy-related signaling pathways, such as BMP and IGF1R/Akt/FOXO (e.g., lncRNAs that inhibit the BMP pathway), may serve as potential therapeutic targets for skeletal muscle atrophy.

2.3.2.5 Driving pyroptosis-mediated muscle atrophy

During the pathological progression of chronic kidney disease (CKD), the expression level of lncRNA GAS5 (lncRNA-GAS5) is significantly upregulated.

Transmission electron microscopy observations reveal that cells exhibit typical morphological characteristics of pyroptosis, which has been identified as a potential pathogenic cell death mode in CKD. Studies have shown that, after knockdown intervention of GAS5, the expression of pyroptosis-related markers (including NLRP3, cleaved caspase-1, and GSDMD-N) is significantly inhibited. Both *in vivo* and *in vitro* experiments demonstrate that the degree of pyroptosis is reduced, accompanied by a significant increase in muscle fiber size. Further verification via RNA pull-down assay and mass spectrometry analysis confirms that GAS5 can interact with the mitochondrial translation elongation factor TUFM. This interaction enhances the activity of NLRP3, thereby inducing pyroptosis and muscle atrophy, while overexpression of TUFM can reverse this effect [74]. Essentially, the interaction between GAS5 and TUFM impairs the anti-pyroptotic function of TUFM, leading to the amplified activation of the NLRP3 signaling pathway, which may be the key mechanism underlying pyroptosis-mediated muscle atrophy.

2.3.2.6 Regulation of miRNA function as competing endogenous RNA

A subset of lncRNAs can serve as competing endogenous RNAs (ceRNAs). By sequestering specific miRNAs, they relieve the inhibitory effects of these miRNAs on their target genes. For example, lncMUMA regulates miRNA molecules via a sponging effect, which reverses skeletal muscle atrophy induced by mechanical unloading and provides a new direction for interventional therapy [75]. Specifically, lncMUMA promoted myogenic differentiation by functioning as a miR-762 sponge to regulate the core myogenic regulator MyoD *in vitro*. The enforced expression of lncMUMA relieved the decreases in MyoD protein and muscle mass in miR-762 knock-in mice. Therapeutically, the enforced expression of lncMUMA improved the *in vitro* myogenic differentiation of myoblasts under microgravity simulation, prevented the development of muscle atrophy, and reversed the established muscle atrophy in hindlimb suspension (HLS) mice.

2.3.2.7 As a repressor of muscle development

SISL (SYNPO2 intronic sense-overlapping lncRNA) is an intronic lncRNA highly expressed in muscle. Studies have demonstrated that SISL functionally promotes myoblast proliferation and fusion while inhibiting myogenic differentiation. SYISL knockout in mice leads to a significant increase in muscle fiber density and muscle mass. Mechanistically, SYISL recruits the enhancer of zeste homolog 2 (EZH2) protein – a core component of the polycomb repressive complex 2 (PRC2) – to the promoters of the cell cycle inhibitory gene p21 and muscle-specific genes (such as myogenin [MyoG], muscle creatine kinase [MCK], and myosin heavy chain 4 [Myh4]). This recruitment results in H3K27 trimethylation and epigenetic silencing of the target genes. These results indicate that SYISL is an inhibitor of muscle development and plays a crucial role in PRC2-mediated myogenesis [76].

2.3.3 CircRNAs in skeletal muscle atrophy

The circRNAs) form covalently closed structures through the back-splicing of precursor mRNAs (pre-mRNAs), and they are characterized by high stability and strong tissue specificity. Their mechanisms of action in skeletal muscle atrophy mainly involve two pathways: encoding functional proteins and constructing ceRNA regulatory networks.

2.3.3.1 Promoting atrophy effect mediated by encoded proteins

The circDdb1 is an identified key pro-atrophic circRNA. Its expression is significantly upregulated in various muscle atrophy models, such as denervation, immobilization, aging, and diabetes, and it is also elevated in human aged muscle tissues. Studies have confirmed that circDdb1 can initiate translation through the internal ribosomal entry site (IRES) to generate a novel protein, CircDdb1-867aa, which contains 867 amino acids. This protein can bind to eukaryotic elongation factor 2 (eEF2) and promote the phosphorylation of eEF2 at Thr56, directly inhibiting the protein translation process. Eventually, this leads to myotube atrophy and the upregulation of atrophy markers such as Atrogin-1 and MuRF1. Further studies have found that eukaryotic initiation factor 4A3 (EIF4A3) can bind to the flanking sequences of circDdb1 to promote its cyclization, forming the "EIF4A3-circDdb1-CircDdb1-867aa" pro-atrophic regulatory axis [77].

The circTmeff1 has been identified as a potential circRNA candidate for targeting muscle atrophy [78]. Studies have shown that the expression level of circTmeff1 is closely associated with muscle atrophy. Its overexpression can induce muscle atrophy in both *in vitro* and *in vivo* models, while knocking down its expression can rescue this pathological process in the same models. At the mechanistic level, circTmeff1 can directly interact with TAR DNA-binding protein 43 (TDP-43), promoting the aggregation of TDP-43 in mitochondria. This further leads to the release of mitochondrial DNA (mtDNA) into the cytoplasm and activates the cyclic GMP-AMP synthase (cGAS)/stimulator of interferon genes (STING) pathway. Meanwhile, circTmeff1 can encode a novel protein, TMEFF1-339aa, which is a key mediator for circTmeff1 to exert its pro-atrophic effects on muscle.

2.3.3.2 Regulatory role of cerna network

In diabetes-related skeletal muscle atrophy, circRNAs participate in the pathological process by constructing a "circRNA-miRNA-mRNA" ceRNA regulatory network [79]. Studies on streptozotocin-induced type 1 diabetes mouse models have shown that there are 1,403 differentially expressed circRNAs in the gastrocnemius muscle, among which 690 are upregulated and 713 are downregulated. The parent genes of these differential circRNAs are significantly enriched in key pathways related to muscle atrophy, such as ubiquitin-mediated protein degradation and the FoxO signaling pathway. Bioinformatics analysis confirms that some differentially expressed circRNAs can act as "molecular sponges" to competitively adsorb miRNAs targeting the atrophy marker MuRF1. Through this competitive regulation mechanism, they affect the expression level of MuRF1, thereby regulating the degradation process of muscle fibers.

2.3.3.3 Mediation of muscle atrophy and myogenesis via interaction with miRNAs

In diverse biological systems, circRNAs have emerged as key regulators of gene expression. Chicken-derived circRNA (circAGO3) is derived from exons 5, 6, 7, and 8 of the AGO3 gene. It exhibits high stability and shows differential expression during the embryonic and post-hatching stages of chickens. Overexpression of circAGO3 in chicken skeletal muscle significantly impairs myogenesis by downregulating muscle differentiation marker genes and upregulating muscle atrophy-related genes. Further analysis combining RNA sequencing and functional assays revealed that circAGO3 can regulate the NF-κB signaling pathway through its interaction with miR-34b-5p, thereby participating in the regulation of inflammatory responses [80].

In addition, the circRNA circCCDC91 can regulate the expression of insulin receptor substrate 1 (IRS1) by sponging three miRNAs, namely miR-15a, miR-15b-5p, and miR-15c-5p, thereby activating the insulin-like growth factor 1-phosphatidylinositol 3-kinase/AKT (IGF1-PI3K/AKT) signaling pathway. Moreover, this circRNA can rescue skeletal muscle atrophy by activating the aforementioned signaling pathway. Specifically, circCCDC91 exerts dual effects of promoting myoblast proliferation and differentiation, as well as alleviating skeletal muscle atrophy, through direct binding to the miR-15 family and activation of the IGF1-PI3K/AKT signaling pathway in chickens [81].

2.3.3.4 Involvement in myogenesis

Spermine oxidase (SMOX), a key polyamine catabolic enzyme, plays an important regulatory role in muscle differentiation during myogenesis. Recent studies have revealed that the SMOX gene can be transcribed to generate circRNA (circSMOX). Experimental results showed that the expression trends of linear SMOX and circSMOX were consistent during the differentiation of C2C12 myoblasts. Moreover, the expression level of circSMOX in the *in vitro* myotube atrophy model and the *in vivo* ALS mouse model was significantly higher than that in the physiological state. This study confirms that SMOX is a novel regulatory factor involved in muscle physiological and pathological processes and provides a scientific basis for subsequent research on circSMOX as a potential target for muscle atrophy therapy [82].

Collectively, ncRNAs form complex regulatory networks that play a crucial role in shaping the pathological processes of muscle diseases. The dysregulation of ncRNAs is both a cause and a consequence of muscle dysfunction, a feature that makes them promising targets for precision therapy and ideal candidates for noninvasive biomarkers. In the future, by integrating high-throughput sequencing technology with functional verification experiments to deeply dissect the regulatory networks of ncRNAs, it will provide key support for the development of novel diagnostic biomarkers and the construction of targeted therapeutic strategies for muscle atrophy-related diseases.

3. The strategies of RNA therapy for muscle disease

The field of RNA biology has been extensively studied since the discovery of mRNA in the 1960s [83]. However, the most significant application of RNA-based therapies emerged during the COVID-19 pandemic. This crisis prompted an unparalleled collaboration between academic institutions and biotechnology companies, accelerating the development of effective counter measures. By December 2020, two mRNA-based vaccines (Moderna and Pfizer/BioNTech) had successfully completed their clinical trials and were deployed on a large-scale application for the prevention of COVID-19 infection [84]. These vaccines demonstrated remarkable efficacy in generating robust and long-lasting antibody responses [85]. Due to this special event, the developed RNA technology not only improved the vaccine industry but also boosted the field of RNA therapy, including protein replacement, genome editing, cell reprogramming, and cell therapy [86].

3.1 Types of RNA-based therapeutics cargo

RNA-based therapeutics have emerged as a promising next-generation medicine, leveraging various types of RNA cargo to address a wide range of clinical needs. These RNA molecules can be broadly categorized based on their size, structure, and function, with significant implications for their therapeutic applications and production challenges. Large RNA molecules, such as mRNA, lncRNA, and certain circRNAs, are typically involved in protein synthesis or the regulation of gene expression. In contrast, small RNA molecules, including small interfering RNA (siRNA), antisense oligonucleotides (ASOs), and miRNA, primarily function through gene silencing and regulation (see **Figure 1**). Additionally, the production of RNA cargoes varies in complexity and cost. Smaller RNA molecules are generally easier and more cost-effective to design and synthesize. However, the production of large RNA molecules is more expensive, although these molecules offer a broader range of therapeutic strategies for RNA therapy.

3.2 Small RNA cargo for gene silencing in muscle therapy

Small-sized oligonucleotides represent the most well-studied and clinically advanced category of RNA-based therapeutics, with several drugs already approved by the FDA. These include ASOs and siRNAs, which primarily function through gene silencing mechanisms [87, 88]. This section will provide an overview of different types of small RNA molecules, highlighting their therapeutic applications and the significance of RNA silencing in muscle diseases.

3.2.1 RNA interference

RNA interference (RNAi) has revolutionized molecular biology since its discovery in 1998 [89], earning the Nobel Prize in 2006 [90]. The RNAi mechanism involves siRNA or short hairpin RNA (shRNA) entering the RNA-induced silencing complex (RISC), which unwinds double-stranded siRNA and guides it to complementary target RNAs. This mechanism allows selective knockdown of disease-associated genes, providing a powerful therapeutic strategy [91]. In addition, chemical modifications, such as 2′-O-Me or 2′-F, enhance RNA stability and reduce immunogenicity, further improving the clinical value of RNAi therapy [92–94].

In muscle disease, the systemic delivery of cholesterol-conjugated siRNA targeting myostatin (Mstn) was developed to achieve muscle-specific gene silencing. With a single intravenous injection of the siRNA, 85%–95% knockdown of Mstn mRNA in various muscles and a >65% reduction in circulating Mstn protein levels were reported in a mouse model over 21 days. This knockdown can lead to significant increases in muscle size, weight, and strength, with no observed toxicity [95]. Similarly, chemically modified siRNAs targeting Alk4 in mdx and wild-type mice

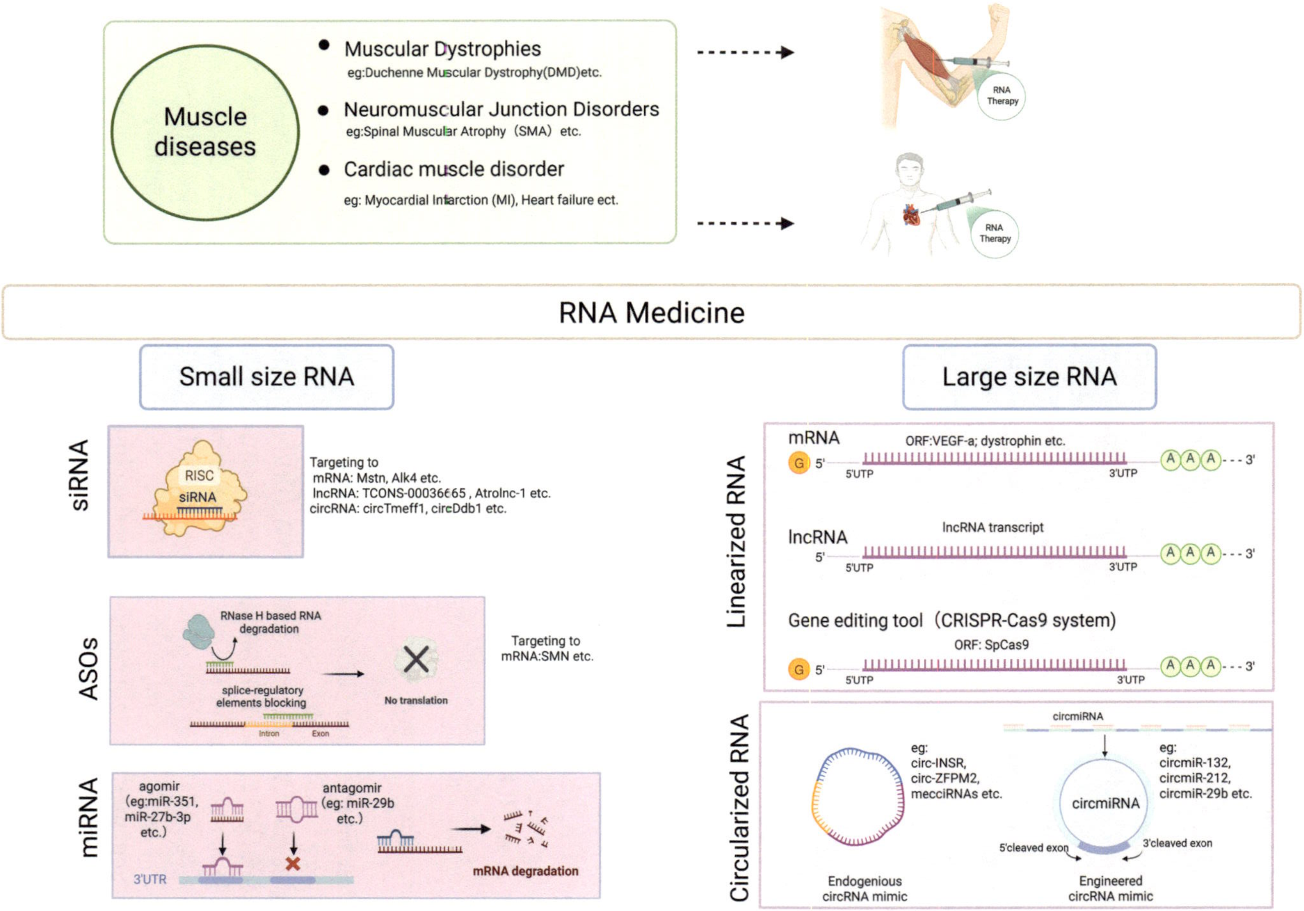

Figure 1.
Schematic graph of multi types of RNA molecules in the therapy of muscle disease (created in https://biorender.com, account from Dongchao Lu).

reduced Alk4 mRNA expression, showing promising therapeutic effects for DMD. Intraperitoneal injection of 10 mg/kg siRNA led to a significant reduction of Alk4 mRNA and clearly improved muscle quality while delaying disease progression [96].

The siRNA is also an effective approach to perform ncRNA inhibition for novel RNA-based therapy for muscle diseases. For example, TCONS-00036665 is a lincRNA that promotes proliferation and inhibits differentiation of pig skeletal muscle satellite cells by interacting with EZH2. Overexpression of TCONS-00036665 led to reduced muscle weight and cross-sectional area of muscle fibers, indicating its inhibitory effect on muscle development. By using the siRNA approach, the repression of TCONS-00036665 can improve muscle cell differentiation as well as maintain muscle function [97]. In addition, Atrolnc-1 is reported to significantly promote muscle wasting in CKD by enhancing protein degradation. Using AAV-shRNA-based therapy targeting Atrolnc-1, the muscle-wasting phenotype was effectively reduced by 42% through suppression of NF-κB activity and MuRF-1 expression in a mouse model [98]. Additionally, the siRNA system could also target special structural RNAs such as circRNAs. Chen et al. [78] reported that circTmeff1 is highly expressed in multiple types of muscle atrophy *in vivo* and *in vitro*. Inhibition of circTmeff1 via the shRNA system could rescue muscle mass in atrophic mice. Similarly, circDdb1 is also a pro-atrophy circRNA, while repression of circDdb1 via shRNA could reverse denervation, angiotensin II, and immobility-induced muscle cell dysfunction [77].

RNAi using siRNA and shRNA has shown significant therapeutic potential in muscle diseases. This effective inhibition system can modulate any kind of RNA molecules, including mRNA, ncRNAs, and circRNAs, providing a good therapeutic strategy for RNA therapy in muscle diseases.

3.2.2 Antisense RNAs

Similar to the siRNA system, ASOs are highly adaptable molecules that can be engineered to selectively bind and modulate RNA transcripts, thereby decelerating or arresting the uncontrolled disease gene expression. For example, Gapmer antisense oligonucleotides (gapmer ASOs) are single-stranded antisense oligonucleotides with a DNA core flanked by RNA-based modifications resistant to RNase H. They target pre-mRNAs and mRNAs, forming DNA:RNA hybrids that recruit RNase H, leading to mRNA degradation and reduced gene expression [99]. In addition, splice-switching antisense oligonucleotides (ssASOs) can be utilized to modulate pre-mRNA splicing by blocking splice-regulatory elements, thereby preventing their recognition by the splicing machinery [100].

In the context of muscle therapy, SMA is frequently caused by deletions or mutations in the survival motor neuron (SMN) gene, leading to severe muscle weakness and significant infant mortality [101–103]. One of the most promising treatments for SMA is the application of ASOs to correct the splicing of exon 7 in the SMN2 gene [104]. After over a decade of research, Spinraza™ became the first FDA-approved treatment for SMA and the first antisense drug to restore the expression of a fully functional protein through splicing correction [105]. The success of ASOs-based RNA therapy in treating SMA highlights the therapeutic potential of modulating RNA transcripts to correct genetic defects. This breakthrough underscores the promise of antisense technology in addressing other muscular disorders, suggesting a bright future for RNA-based therapies in muscle disease management.

3.2.3 Micro-RNAs

In miRNA-based therapies for muscle diseases, the key challenges are to inhibit detrimental miRNAs and promote beneficial ones, thereby developing effective therapeutic strategies.

Two distinct strategies for miRNA silencing are antagomirs and miRNA sponges. Antagomirs are chemically modified oligonucleotides designed to bind specifically to individual miRNAs [106]. The miRNA sponges, on the contrary, are transcripts containing multiple tandem binding sites for a specific miRNA of interest [107]. For instance, miR-29b has been identified as a key modulator of muscle atrophy in various muscle disorders. Both miR-29b antagomirs and miR-29b sponges have been used to effectively reduce elevated miR-29b levels in different mouse models of muscle atrophy, resulting in significant functional recovery of muscle cells [60, 108]. Antagomirs can be further modified and screened to improve their pharmacological effectiveness. Liu et al. [109] screened for a novel miR-29b inhibitor by considering both the three-dimensional structure of pre-miR-29b and the thermodynamics of its interaction with small molecules. They found that TGP-29b-066 exhibited superior miR-29b inhibition efficiency and therapeutic efficacy for muscle atrophy compared to commercial miR-29b inhibitors [109].

To mimic the structure and function of endogenous miRNAs, agomir is developed as a chemically modified and synthetic RNA molecule for the application of miRNA overexpression therapy. The chemical modifications of agomir can enhance the stability and efficiency of small molecules, making them valuable tools for studying gene regulation and exploring therapeutic applications [110]. For example, miR-351 was repressed in the TA muscle at various times following sciatic nerve transection stress. He et al. [111] reported using miR-351 agomir as the overexpression strategy to restore the miR-351 expression under disease stress. They observed that the chemically modified miR-351 mimic inhibited denervation-induced TA muscle atrophy and improved muscle function *in vivo* [111]. Additionally, Yao et al. reported that miR-27b-3p modulates the fibrogenic differentiation of fibro-adipogenic progenitors. Agomir-27b-3p effectively suppressed fibrosis in both *in vitro* and *in vivo* models of denervation and rescued fibrosis in chronic skeletal muscle injury [112].

The therapeutic potential of miRNAs in muscle diseases is evident from the successful application of antagomirs, miRNA sponges, and agomirs in modulating miRNA activity. These approaches have shown significant promise in reducing muscle atrophy and improving muscle function in various models. The continued advancement in miRNA technologies holds great promise for addressing unmet needs in muscle disease treatment.

3.3 Large size of RNA cargo for muscle therapy

Large-sized oligonucleotides are typically *in vitro*-transcribed RNA molecules with a size of over 200nt in length. Any RNA molecule can be synthesized through *in vitro* transcription (IVT) using T7/SP6 RNA polymerases in a cell-free system [113]. For example, modified mRNA (modRNA), which consists of a 5′-cap, poly-A tail, 5′- and 3′- UTRs, and an open reading frame (ORF), can be prepared via IVT. A linearized DNA template containing a T7 promoter, the gene of interest, and a poly-A sequence is used as the IVT template, followed by T7 RNA polymerase-based IVT and the incorporation of modified nucleotides like N1-methylpseudouridine to enhance stability. In recent years, the development of mRNA drugs has seen the application of three distinct types of mRNAs: non-replicating mRNAs (nrRNAs), self-amplifying mRNAs (saRNAs), and circRNAs. The nrRNAs are the most straightforward, encoding only the protein of interest and offering ease of production and scalability. The saRNAs incorporate viral replicase genes, enabling intracellular amplification and thus requiring lower doses to achieve significant protein expression [114]. Circular mRNAs, formed by joining the 5′ and 3′ ends, are highly stable due to their resistance to exonucleases, resulting in longer half-lives and sustained protein production [115]. Each type has unique advantages and challenges, influencing their suitability for different therapeutic applications. In addition to mRNAs, ncRNA medicines can also be generated via IVT with both linearized and circularized structures. The diverse types of large RNA cargo hold great potential for novel RNA therapies. In this section, we provide an overview of the applications of large RNA molecules in muscle diseases.

3.3.1 Linearized RNA cargo

Since 1990, the concept of nucleic acid-encoded drugs has been explored by directly injecting IVT mRNA into the skeletal muscle of mice to evaluate protein expression [116]. To date, IVT RNA has emerged as a promising new drug class for delivering genetic information. These synthetic RNAs can be engineered to transiently express proteins by mimicking the structure of natural mRNA, express ncRNA transcripts, or even encode specialized gene-editing molecules such as Cas9 mRNA [117].

In the context of muscle disease, transiently secreted vascular endothelial growth factor (VEGF) has demonstrated beneficial effects on myocardial infarction-mediated cardiomyocyte dysfunction. ModRNA encoding VEGF-A has been reported as a promising protein replacement therapy for cardiac muscle repair. Specifically, transfection of VEGF-A modRNA into iPSC-derived cardiomyocytes (iPSC-CMs) enhances their survival rates and facilitates the formation of a dense vascular network during iPSC-CM transplantation therapy [118]. Similarly, in larger animal models, VEGF-A modRNA has been shown to improve heart function and promote angiogenesis [119]. Additionally, for the DMD therapy, a novel approach using Bobcat mRNA™ has been developed. This proprietary mRNA, encoding a full-length human dystrophin protein (mRNA-DMD), has demonstrated the ability to restore muscle strength in D2.mdx mutant mice. Furthermore, a Phase I clinical trial assessing the safety and tolerability of full-length dystrophin mRNA is being conducted (NCT07188012). The successful development and delivery of mRNA-DMD offer a promising alternative or complementary approach to existing therapies for DMD.

It is important to note that some muscular dystrophies are genetic diseases that induce muscle degeneration [120]. Correcting the mutated genes and repairing muscle function are key challenges in this field. CRISPR/Cas9 gene correction in transplantable cells offers hope for restoring organ function. However, muscle cells have been slow to benefit from this advanced technology due to difficulties in *ex vivo* or *in vivo* manipulation. Stadelmann et al. [121] developed an mRNA-based delivery system for SpCas9 and an adenine base editor to perform genome editing in human muscle stem cells. They demonstrated that cells edited with this mRNA-based gene editing system can fully retain their myogenic marker signature, proliferation capacity, and functional attributes. Moreover, this mRNA-based delivery of a base editor efficiently repaired a muscular dystrophy-causing SGCA mutation in a single, selection-free step.

The development of linearized RNA cargo, including IVT mRNA or RNA-based gene editing systems, has shown significant promise in addressing muscle degeneration and genetic defects in muscular dystrophies. Future research should focus on optimizing delivery systems and exploring the therapeutic potential of these RNA-based approaches in clinical settings. The continued innovation in RNA cargo design and delivery holds great promise for advancing the field of muscle disease therapy.

3.3.2 Circularized RNA medicine

The circRNAs represent a class of covalently closed, single-stranded RNA molecules identified across various species and implicated in disease regulation. Their distinctive stability arises from the back-splicing process, wherein the 3′-end of an exon links to the 5′-end of an upstream exon through a 3′,5′-phosphodiester bond. This structure leads to resistance to RNase R degradation and extends the half-life of RNA molecules, thereby enhancing their functional capacity [122]. For therapy in muscle disease, *in vitro* transcribed circRNA (ivcRNA) cargo has also shown potential therapeutic value in recovering muscle cell functions. For example, circ-INSR is one well-known circRNA that serves as a key regulator of muscle development. Silencing of circ-INSR reduces the proliferation and increases apoptosis of myocytes, while the overexpression of circ-INSR can promote bovine primary myocyte function [123]. In cardiac muscle cells, circ-INSR was also observed as an important mediator in maintaining cardiomyocyte function under doxorubicin stress. By using an IVT system as well as the T4-ligase-based circularization approach, circ-INSR-based RNA medicine has been generated. This circRNA medicine demonstrated a strong protective effect against doxorubicin-induced cell death in both rodent and human cardiomyocytes [124]. Furthermore, using the same production method, Neufeldt et al. [125] generated recombinant circZFPM2 RNA and further proved that this circRNA medicine can reduce dysfunction of cardiomyocytes under doxorubicin stress. In addition, Liu et al. reported that *in vitro* transcribed mitochondria-encoded circular RNAs (mecciRNAs) are also functional circRNA medicines that can repair cardiomyocyte function. Notably, IVT-mecciRNAs demonstrated a clear cardiomyocyte protective effect in both doxorubicin and pressure overload–induced heart failure mouse models, indicating a good translational value of circRNA medicine for disease therapy.

Endogenous circRNAs demonstrate therapeutic potential not only because of their target genes but also due to their unique structure, which enhances stability and functional efficacy. Thus, the circularized structure has been further utilized to generate the engineered circRNA [126]. For example, Lavenniah et al. [127] applied group I permutated intron–exon sequences to generate the exogenous artificial circRNA sponge system (circmiRs). This circmiRs construct includes 12 miR-132 and -212 sponge binding sites (SBS) with 12 nucleotides spacing between each miRNA

SBS. Notably, this circmiRs medicine demonstrated improved miRNA inhibition efficacy compared to commercial pharmacological antagomiRs and a better therapeutic effect on cardiomyocyte hypertrophy in a mouse heart failure model [127]. In the context of muscle atrophy, Li et al. [128] also designed circmiRs medicine targeting miR-29b. CircmiR-29b effectively inhibits miR-29b in a highly specific and significant manner. It protects against muscle atrophy both *in vitro* and *in vivo*, including denervation, immobilization, and angiotensin II-mediated muscle atrophy models [128].

The development of circularized RNA medicines represents a significant advancement in the field of RNA therapeutics. CircRNAs, with their unique stability and functional efficacy, offer a promising approach for treating muscle and cardiac diseases. Their ability to resist degradation and maintain prolonged activity makes them ideal candidates for addressing conditions like muscle atrophy and cardiac muscle dysfunction.

4. RNA delivery system in muscle disease therapy

Skeletal muscle disorders, such as DMD, have long posed significant therapeutic challenges. Conventional therapies, like corticosteroids, can only delay disease progression and alleviate symptoms but fail to address the underlying genetic defects [129]. While more radical gene therapies hold promise, they are often limited by the packaging capacity, immunogenicity, and potential integration risks of delivery vectors such as AAV, making it difficult to safely and effectively correct large pathogenic genes like DMD [130]. In contrast, RNA-based therapeutics have emerged as a highly promising approach, with the core advantage of circumventing genomic challenges by directly targeting disease-related gene expression processes. Modalities such as ASOs, siRNAs, mRNAs, and circRNAs offer the potential for more sustained and efficient protein expression [77, 131–134].

However, the efficient, safe, and specific delivery of RNA therapeutics to systemic muscle tissues remains the central bottleneck in realizing their therapeutic potential. This process faces multiple inherent physiological and cellular barriers: Instability and susceptibility to degradation of RNA molecules: Naked RNA is rapidly degraded by ubiquitous nucleases in the systemic circulation, with an extremely short half-life, making it difficult for sufficient intact drug to reach the target sites [135]. Difficulty in crossing the vascular endothelial barrier: The dense microstructure of the vascular endothelium in muscle tissues impedes effective extravasation of large molecules and nanoparticles into the interstitial space, thereby limiting drug contact with target muscle cells [136]. Rapid clearance and uptake by non-target organs: Systemically administered delivery vectors, particularly nanoparticles, are easily recognized and cleared by the mononuclear phagocyte system (e.g., liver and spleen), leading to significantly reduced enrichment in target tissues and potential off-target effects such as hepatotoxicity [137]. Low efficiency of cellular internalization and endosomal escape: Even if the drug is endocytosed by muscle cells, most of it becomes trapped in endosomes and is ultimately transported to lysosomes for degradation [138]. Potential immunogenicity: Both exogenous RNA molecules themselves (e.g., if recognized as viral RNA) and certain delivery vector materials may activate the host's innate immune system, triggering inflammatory responses that not only reduce therapeutic efficacy but also pose safety risks [139].

Lipid nanoparticles (LNPs) and engineered extracellular vesicles (EVs), as two of the most promising delivery platforms, can effectively circumvent these challenges and enable efficient and safe RNA delivery.

4.1 LNP-based approaches

LNPs are composed of multiple distinct components. The ionizable lipid, as the core constituent of LNPs, carries a positive charge under acidic pH conditions, enabling it to encapsulate negatively charged RNA. It undergoes a charge reversal within endosomes, thereby promoting endosomal escape. Phospholipids form the LNP bilayer structure, stabilizing the particle and mimicking the cell membrane to facilitate fusion. Cholesterol enhances LNP stability and membrane fluidity, aiding in endosomal escape. PEGylated lipids improve nanoparticle stability and circulation half-life while reducing nonspecific protein binding [140, 141].

LNPs represent one of the most mature nucleic acid delivery systems. Their continued refinement through chemical modifications, ligand conjugation, novel lipid design, and innovative delivery strategies has significantly advanced the clinical translation of circRNA in fields such as cardiovascular diseases, cancer, and genetic disorders. Currently, LNP-encapsulated circRNA RBD-Omicron and circRNA RBD-Delta vaccines can effectively address current SARS-CoV-2 variants by inducing high levels of neutralizing antibodies and a Th1 immune response [142]. LNP-mediated delivery of siRNA targeting the Alzheimer's disease risk locus transcription factor PU.1 successfully reduced PU.1 levels in microglial-like cells and mitigated neuroinflammation in mice injected with LPS and in CK–p25 mice modeling the chronic neuroinflammation common in AD patients [143]. By designing synthetic ivcRNA encoding therapeutic proteins (e.g., MSI2 and SOX5) and encapsulating them using optimized LNPs, direct delivery of ivcRNA-LNPs into the joint cavity of mice effectively delayed cartilage destruction, reduced abnormal bone calcification, and improved histological scores of the joint [144]. Furthermore, for DMD, where repeated AAV injection is limited by neutralizing antibodies, the LNP system enables repeated intramuscular injections to deliver Cas9 mRNA and sgRNA to skeletal muscle. This approach induced stable genomic exon skipping and restored dystrophin expression in a DMD mouse model carrying a humanized exon sequence. Administration of LNPs via limb perfusion can target multiple muscle groups to treat skeletal muscle diseases [134].

To further enhance the muscle tissue targeting of LNPs, large ionizable lipid libraries can be constructed to screen for novel lipids with inherent tropism for muscle tissue. One study proposed a platform for rapidly creating chemically diverse lipids to build LNPs and successfully identified an ionizable lipid, iso-A11B5C1. Incorporating this lipid into LNPs enabled highly efficient muscle-specific mRNA delivery while minimizing off-target delivery to other tissues (e.g., liver and spleen). Although this novel LNP showed limited transfection in lymph nodes, intramuscular injection elicited a robust cellular immune response. An mRNA melanoma vaccine developed using this novel LNP demonstrated potent anti-tumor effects in a mouse melanoma model [145]. Beyond self-composition adjustments, the specificity of LNP delivery to muscle tissue can be directly enhanced by conjugating targeting ligands to the LNP surface or modifying their chemical composition. Research has uncovered the rational design of peptide-ionizable lipids for assembling LNPs with organ-selective mRNA delivery. By incorporating artificial ionizable and natural amino acids and functional molecules into peptide-ionizable lipids, LNPs with tissue-specific targeting capabilities – including to the lung, liver, spleen, thymus, and bone – were created [146]. Another study designed an LNP whose surface displays five specific receptor complements, enabling specific targeting of mouse skeletal muscle with no off-target effects on other tissues and organs [147].

As an RNA delivery platform, LNPs have demonstrated significant clinical application value. However, the targeting specificity for muscle tissue still requires improvement. Concurrently, reducing the incidence of immune responses and mitigating the potential toxicity of repeated administration are crucial to advancing their development for treating muscle diseases.

4.2 Engineered EV-mediated RNA delivery

EVs are membrane-bound vesicles ranging from nanoscale to microscale, actively secreted by cells. They naturally carry important biomolecules such as proteins, lipids, and nucleic acids from their source cells, serving as key messengers in intercellular communication [148]. As delivery vehicles, EVs exhibit unique inherent advantages: their endogenous origin grants them excellent biocompatibility and low immunogenicity, making them less likely to be recognized and cleared by the immune system, thus highly suitable for repeated administration; certain sources of EVs, such as those derived from mesenchymal stem cells, possess a natural homing capability to sites of injury and inflammation, including diseased muscle; as natural mediators of intercellular communication, their membrane structure enables efficient fusion with or internalization by target cells, facilitating effective cargo delivery; furthermore, their nanoscale size gives them significant potential to penetrate the dense extracellular matrix of muscle tissue [149]. Efficiently loading exogenous therapeutic RNA into EVs is a crucial step for realizing their delivery function. Currently, various loading strategies have been developed, each with its own advantages and disadvantages. Co-incubation is the simplest passive loading method, but its efficiency is generally low. Electroporation is one of the most widely used techniques; however, the applied electrical pulses may damage the EV membrane integrity and the structure of the encapsulated RNA. The transfection reagent-based method involves first introducing the RNA into the parent cells that secrete EVs, thereby leveraging the cell's natural machinery to package the RNA into subsequently secreted EVs, achieving indirect loading [150]. Additionally, a series of active loading techniques, such as freeze–thaw cycles, sonication, or the use of permeabilizing agents like saponin, actively load RNA into the lumen by temporarily disrupting EV membrane permeability. These methods offer different trade-offs between loading efficiency and preservation of EV function [151, 152].

Furthermore, EVs can be engineered to enhance their targeting specificity. Using genetic engineering approaches, parent cells can be made to express transmembrane proteins displaying targeting peptides (e.g., muscle-targeting peptides). Pioneering research in an Alzheimer's disease model successfully expressed a Lamp2b protein fused to the neuron-targeting peptide RVG in parent cells. The resulting EVs effectively delivered siRNA to neurons in mouse brains, significantly reducing target gene expression [153]. Similar strategies have been applied to target other tissues, such as using muscle-targeting peptides to modify EVs for enhanced delivery efficiency to muscle. One study identified a peptide, CP05, which binds to the CD63 protein expressed on exosomes. Modifying exosomes by conjugating CP05 to a muscle-targeting peptide increased the delivery of splice-correcting oligonucleotides to muscle, enhancing dystrophin expression and muscle function in a mouse model of muscular dystrophy [154].

Currently, the application of engineered EVs for treating muscle diseases is a rapidly developing research area, with several studies providing proof-of-concept validation. Research has found that inhibiting miR-29b can reduce muscle atrophy. Artificially engineered EVs were constructed to deliver CRISPR/Cas9 targeting miR-29b (EVs-Cas9-29b). EVs-Cas9-29b demonstrated good functional efficacy through specific and rapid inhibition of miR-29b via gene editing. Under both *in vitro* and *in vivo* conditions, EVs-Cas9-29b could prevent muscle atrophy induced by various factors [155]. EVs derived from fibro-adipogenic progenitors, enriched with miR-125b-5p, were able to promote macrophage polarization toward the M2 reparative phenotype by targeting the PTPN1/ERK pathway, improve muscle stem cell activity, accelerate muscle regeneration, and reduce fibrosis [156]. Systemic delivery of muscle-derived EVs was shown to enhance synaptic plasticity and alleviate cognitive impairment induced by chronic cerebral hypoperfusion. Constructing brain-targeted EVs enriched with miR-17/20a-5p, which binds to the DEP domain of DEPTOR, an mTOR-interacting protein, activated the mTOR pathway in the hippocampus and improved cognitive function in a model of vascular cognitive impairment [157].

Despite the great potential shown by engineered EVs in treating muscle atrophy, their clinical translation still faces numerous challenges. Future efforts should focus on developing standardized and scalable production processes to ensure the quality, safety, and batch-to-batch consistency of EV products; actively exploring novel source cells, such as muscle progenitor cells, to obtain EVs with enhanced targeting or functionality; and dedicating efforts to designing smarter "engineered EVs." Through precise modification of their membrane surface or cargo, these EVs could be engineered to respond to specific in vivo microenvironmental signals (e.g., inflammation, ROS), thereby achieving on-demand drug release and ultimately improving the specificity and efficiency of treatment.

5. Clinical trials of RNA therapy in muscle diseases

RNA-therapeutic strategies – particularly exon skipping, AAV-mediated gene delivery, and AOC technology – have demonstrated considerable promise for muscle disorders. The clinical trials summarized in **Table 2** span the major disease categories: DMD, FSHD, DM1, SMA, MG, and LGMD; several agents are now in Phase III or have already been filed for marketing approval. Continued surveillance of long-term efficacy and safety data, together with the rational design of combination regimens, will be critical next steps.

6. Advantages and challenges of RNA therapy in muscle diseases – considerations for future development.

6.1 Advantages of RNA therapy in muscle diseases

More therapeutic targets: Compared with traditional protein-targeting strategies, RNA therapy has more therapeutic targets. RNA therapy can target both

Indication	Clinical trial	Clinical phase	Therapeutic target
Duchenne muscular dystrophy (DMD)	Viltolarsen (Viltepso) [158]	II	Exon 53 of the DMD gene
	WVE-N531 (ClinicalTrials: NCT04906460)	II	Phosphorodiamidate morpholino oligonucleotide (PMO)-modified antisense oligonucleotide targeting exon 53
	Delandistrogene moxeparvovec (SRP-9001) [159]	III	AAV vector delivery of micro-dystrophin gene
Facioscapulohumeral muscular dystrophy (FSHD)	AOC 1020 (Delpacibart Braxlosiran) (ClinicalTrials: NCT03375164)	I/II FORTITUDE trial	Antibody-oligonucleotide conjugates (AOCs) target DUX4 mRNA and inhibit its expression
Myotonic dystrophy type 1 (DM1)	DYNE-101 (Z-basivarsen) (ClinicalTrials: NCT05481879)	I/II ACHIEVE trial	AOC targets DMPK mRNA to correct splicing abnormalities
	VX-670 (ClinicalTrials: NCT06185764)	I/II Galileo trial	Cyclic peptide-conjugated oligodeoxynucleotide targeting abnormal CUG RNA
Spinal muscular atrophy (SMA)	Nusinersen (ClinicalTrials: NCT02386553)	Nurture Study/ Chinese registry study	Antisense oligonucleotides (ASOs) modify SMN2 pre-mRNA splicing to increase full-length SMN protein
	Risdiplam (ClinicalTrials: NCT03779334)	RAINBOWFISH trial	Oral SMN2 splicing modifier to increase systemic SMN protein levels
Myasthenia gravis	Cemdisiran (ClinicalTrials: NCT05070858)	III	Small interfering RNA (siRNA) drug targeting complement factor 5 (C5)
Limb-girdle muscular dystrophy (LGMD)	Bidridistrogene xeboparvovec (SRP-9,003) (ClinicalTrials: NCT0,6246,513)	III SRP-9003-301 trial	AAV vector delivers the SGCB gene to treat LGMD2E/R4

Table 2.
The clinical trials of skeletal muscle diseases.

protein-coding RNAs and ncRNAs. Even more, RNA therapy can intervene at the post-transcriptional RNA modification level. This provides more options for the treatment targets of muscle diseases.

Rapid and flexible drug design: RNA therapy drugs (such as mRNA and siRNA) can directly encode functions through base sequences without the need to fold complex three-dimensional structures like protein drugs. Once the target sequence is determined, corresponding RNA drugs can be synthesized within a few days (such as COVID-19 mRNA vaccines). In addition, AI-assisted design platforms enable batch and personalized drug design strategies, greatly accelerating the drug development cycle.

Long-lasting effects: Some RNA therapies (such as Inclisiran, PCSK9 siRNA) can provide a durable effect. This is beneficial in chronic muscle diseases where sustained therapeutic effects are desired, which can also improve patient compliance.

6.2 Challenges in RNA therapy for muscle diseases

Delivery systems: Efficient delivery of RNA molecules to muscle remains a major hurdle. Muscle-specific delivery systems still need to be developed. LNPs, exosomes, AAV virus, or others all need to improve the muscle-targeted delivery strategy.

Stability and immunogenicity: Although multiple chemical modifications have been used to improve stability and reduce immunogenicity, these two aspects remain the main factors affecting the effectiveness and safety of RNA therapy.

AI-assisted design platforms: The AI-assisted RNA drug design platform has brought new hope for the treatment of these diseases. However, designing and developing reliable AI platforms also seem to be an urgent matter.

7. Conclusion

RNA therapy holds transformative potential for treating muscle diseases. Continued development in delivery technologies, RNA modification strategies, and AI-assisted design platforms will advance RNA-based treatments for muscle diseases in the future.

Acknowledgments

This work was supported by grants from the Shanghai Pujiang Program (23PJ1,412200 to D.L.). The graphic abstract was created with BioRender.com.

Conflict of Interest

The authors declare no conflict of interest.

Declarations

None.

DOI: http://dx.doi.org/10.5772/intechopen.1014065

Author details

Dongchao Lu[1] and Jin Li[2*]

1 School of Integrative Medicine, Shanghai University of Traditional Chinese Medicine, Shanghai, China

2 Institute of Cardiovascular Sciences, Shanghai Engineering Research Center of Organ Repair, Joint International Research Laboratory of Biomaterials and Biotechnology in Organ Repair (Ministry of Education), School of Life Sciences, Shanghai University, Shanghai, China

*Address all correspondence to: ahulijin@shu.edu.cn

References

[1] TA J, TL J, Kääriäinen M, Kalimo H, Järvinen M. Muscle injuries: Biology and treatment. The American Journal of Sports Medicine. 2005;**33**(5):745–764. DOI: 10.1177/0363546505274714

[2] Latronico N, Bolton CF. Critical illness polyneuropathy and myopathy: A major cause of muscle weakness and paralysis. The Lancet Neurology. 2011;**10**(10):931–941. DOI: 10.1016/s1474-4422(11)70178-8

[3] Mammen AL. Statin-associated autoimmune myopathy. The New England Journal of Medicine. 2016;**374**(7):664–669. DOI: 10.1056/NEJMra1515161

[4] Leone E, Pandyan A, Rogers A, Kulshrestha R, Hill J, Philp F. Effectiveness of conservative non-pharmacological interventions in people with muscular dystrophies: A systematic review and meta-analysis. Journal of Neurology, Neurosurgery, and Psychiatry. 2024;**95**(5):442–453. DOI: 10.1136/jnnp-2023-331988

[5] Huang M, Magni N, Rice D. The prevalence, characteristics and impact of chronic pain in people with muscular dystrophies: A systematic review and meta-analysis. The Journal of Pain. 2021;**22**(11):1343–1359. DOI: 10.1016/j.jpain.2021.04.001

[6] Paik J. Risdiplam: A review in spinal muscular atrophy. CNS Drugs. 2022;**36**(4):401–410. DOI: 10.1007/s40263-022-00910-8

[7] Finkel RS, Hughes SH, Parker J, Civitello M, Lavado A, Mefford HC, Mueller L, Kletzl H. Risdiplam for prenatal therapy of spinal muscular atrophy. The New England Journal of Medicine. 2025;**392**(11):1138–1140. DOI: 10.1056/NEJMc2300802

[8] Kino Y, Washizu C, Oma Y, Onishi H, Nezu Y, Sasagawa N, Nukina N, Ishiura S. MBNL and CELF proteins regulate alternative splicing of the skeletal muscle chloride channel CLCN1. Nucleic Acids Research. 2009;**37**(19):6477–6490. DOI: 10.1093/nar/gkp681

[9] Ciaranello AL, Leroy V, Rusibamayila A, Freedberg KA, Shapiro R, Engelsmann B, Lockman S, Kelly KA, Dabis F, Walensky RP. Individualizing the WHO HIV and infant feeding guidelines: Optimal breastfeeding duration to maximize infant HIV-free survival. AIDS (London, England). 2014;**3**(03):S287–299. DOI: 10.1097/qad.0000000000000337

[10] Nakka K, Ghigna C, Gabellini D, Dilworth FJ. Diversification of the muscle proteome through alternative splicing. Skeletal muscle. 2018;**8**(1):8. DOI: 10.1186/s13395-018-0152-3

[11] Childs-Disney JL, Stepniak-Konieczna E, Tran T, Yildirim I, Park H, Chen CZ, Hoskins J, Southall N, Marugan JJ, Patnaik S, Zheng W, Austin CP, Schatz GC, Sobczak K, Thornton CA, Disney MD. Induction and reversal of myotonic dystrophy type 1 pre-mRNA splicing defects by small molecules. Nature Communications. 2013;**4**:2044. DOI: 10.1038/ncomms3044

[12] Tanner MK, Tang Z, Thornton CA. Targeted splice sequencing reveals RNA toxicity and therapeutic response in

myotonic dystrophy. Nucleic Acids Research. 2021;**49**(4):2240–2254. DOI: 10.1093/nar/gkab022

[13] Goodwin M, Mohan A, Batra R, Lee K-Y, Charizanis K, Fernández Gómez FJ, Eddarkaoui S, Sergeant N, Buée L, Kimura T, Clark HB, Dalton J, Takamura K, Weyn-Vanhentenryck SM, Zhang C, Reid T, Ranum LW, Day JW, Swanson MS. MBNL sequestration by toxic RNAs and RNA misprocessing in the myotonic dystrophy brain. Cell Reports. 2015;**12**(7):1159–1168. DOI: 10.1016/j.celrep.2015.07.029

[14] Wang G-S, Kearney DL, De Biasi M, Taffet G, Cooper TA. Elevation of RNA-binding protein CUGBP1 is an early event in an inducible heart-specific mouse model of myotonic dystrophy. The Journal of Clinical Investigation. 2007;**117**(10):2802–2811. DOI: 10.1172/jci32308

[15] Masuda A, Andersen HS, Doktor TK, Okamoto T, Ito M, Andresen BS, Ohno K. CUGBP1 and MBNL1 preferentially bind to 3′ UTRs and facilitate mRNA decay. Scientific Reports. 2012;**2**:209. DOI: 10.1038/srep00209

[16] Tao Y, Zhang Q, Wang H, Yang X, Mu H. Alternative splicing and related RNA binding proteins in human health and disease. Signal Transduction and Targeted Therapy. 2024;**9**(1):26. DOI: 10.1038/s41392-024-01734-2

[17] Sidlauskaite E, Le Gall L, Mariot V, Dumonceaux J.DUX4 expression in FSHD muscles: Focus on its mRNA regulation. Journal of Personalized Medicine. 2020;**10**(3). DOI: 10.3390/jpm10030073

[18] García-Rodríguez R, Hiller M, Jiménez-Gracia L, van der Pal Z, Balog J, Adamzek K, Aartsma-Rus A, Spitali P (2020) Premature termination codons in the DMD gene cause reduced local mRNA synthesis. Proceedings of the National Academy of Sciences of the United States of America **117**:16456–16464. DOI: 10.1073/pnas.1910456117

[19] Leckie J, Zia A, Yokota T.An updated analysis of exon-skipping applicability for duchenne muscular dystrophy using the UMD-DMD database. Genes. 2024;**15**(11). DOI: 10.3390/genes15111489

[20] Hildebrandt RP, Moss KR, Janusz-Kaminska A, Knudson LA, Denes LT, Saxena T, Boggupalli DP, Li Z, Lin K, Bassell GJ, Wang ET. Muscleblind-like proteins use modular domains to localize RNAs by riding kinesins and docking to membranes. Nature Communications. 2023;**14**(1):3427. DOI: 10.1038/s41467-023-38923-6

[21] Alvarez K, Polanco-Roman L, Samuel Breslow A, Molock S. Structural racism and suicide prevention for ethnoracially minoritized youth: A conceptual framework and illustration across systems. The American Journal of Psychiatry. 2022;**179**(6):422–433. DOI: 10.1176/appi.ajp.21101001

[22] Zhang H, Hu Z, Wang S, Xu J, Li S, Song X. Association of general and abdominal obesity with lung function, FeNO, and blood eosinophils in adult asthmatics: Findings from NHANES 2007-2012. Frontiers in Physiology. 2023;**14**:1019123. DOI: 10.3389/fphys.2023.1019123

[23] Iwasaki S, Shojaku H, Kawahara T, Shindo S, Miyashita T, Aoki M, Tsubota M, Takeda N. Estimated

prevalence and characteristics of bilateral vestibulopathy diagnosed in Japan: A nationwide survey. Auris, Nasus, Larynx. 2022;**49**(3):347–351. DOI: 10.1016/j.anl.2021.08.008

[24] Lin X, Zhou Y, Xue L. Mitochondrial complex I subunit MT-ND1 mutations affect disease progression. Heliyon. 2024;**10**(7): e28808. DOI: 10.1016/j.heliyon.2024. e28808

[25] Ruzzenente B, Metodiev MD, Wredenberg A, Bratic A, Park CB, Cámara Y, Milenkovic D, Zickermann V, Wibom R, Hultenby K, Erdjument-Bromage H, Tempst P, Brandt U, Stewart JB, Gustafsson CM, Larsson NG. LRPPRC is necessary for polyadenylation and coordination of translation of mitochondrial mRNAs. The EMBO Journal. 2012;**31**(2):443–456. DOI: 10.1038/emboj.2011.392

[26] Coenen MJ, Antonicka H, Ugalde C, Sasarman F, Rossi R, Heister JG, Newbold RF, Trijbels FJ, van den Heuvel LP, Shoubridge EA, Smeitink JA. Mutant mitochondrial elongation factor G1 and combined oxidative phosphorylation deficiency. The New England Journal of Medicine. 2004;**351**(20):2080–2086. DOI: 10.1056/NEJMoa041878

[27] Yang H, Wang M, Liu X, Zhao XM, Li A. PhosIDN: An integrated deep neural network for improving protein phosphorylation site prediction by combining sequence and protein-protein interaction information. Bioinformatics (Oxford, England). 2021;**37**(24):4668–4676. DOI: 10.1093/bioinformatics/btab551

[28] Demeke HB, McCray E, Dean HD. Ending the HIV Epidemic in the United States. The American Journal of Nursing. 2020;**120**(3):21–22. DOI: 10.1097/01.NAJ.0000656304.84951.da

[29] Roundtree IA, Evans ME, Pan T, He C. Dynamic RNA modifications in gene expression regulation. Cell. 2017;**169**(7):1187–1200. DOI: 10.1016/j.cell.2017.05.045

[30] Delaunay S, Helm M, Frye M. RNA modifications in physiology and disease: Towards clinical applications. Nature Reviews Genetics. 2024;**25**(2):104–122. DOI: 10.1038/s41576-023-00645-2

[31] Gatsiou A, Stellos K. RNA modifications in cardiovascular health and disease. Nature Reviews Cardiology. 2023;**20**(5):325–346. DOI: 10.1038/s41569-022-00804-8

[32] Sun J, Zhou H, Chen Z, Zhang H, Cao Y, Yao X, Chen X, Liu B, Gao Z, Shen Y, Qi L, Sun H. Altered m6A RNA methylation governs denervation-induced muscle atrophy by regulating ubiquitin proteasome pathway. Journal of Translational Medicine. 2023;**21**(1):845. DOI: 10.1186/s12967-023-04694-3

[33] Widmann M, Nieß AM, Munz B. Physical exercise and epigenetic modifications in skeletal muscle. Sports Medicine (Auckland, NZ). 2019;**49**(4):509–523. DOI: 10.1007/s40279-019-01070-4

[34] Li J, Pei Y, Zhou R, Tang Z, Yang Y. Regulation of RNA N(6)-methyladenosine modification and its emerging roles in skeletal muscle development. International Journal of Biological Sciences. 2021;**17**(7): 1682–1692. DOI: 10.7150/ijbs.56251

[35] Fu Y, Dominissini D, Rechavi G, He C. Gene expression regulation mediated through reversible m(6)A

RNA methylation. Nature Reviews Genetics. 2014;**15**(5):293–306. DOI: 10.1038/nrg3724

[36] Zaccara S, Ries RJ, Jaffrey SR. Reading, writing and erasing mRNA methylation. Nature Reviews Molecular Cell Biology. 2019;**20**(10):608–624. DOI: 10.1038/s41580-019-0168-5

[37] Boulias K, Greer EL. Biological roles of adenine methylation in RNA. Nature Reviews Genetics. 2023;**24**(3):143–160. DOI: 10.1038/s41576-022-00534-0

[38] Shi H, Wei J, He C. Where, when, and how: Context-dependent functions of RNA methylation writers, readers, and erasers. Molecular Cell. 2019;**74**(4):640–650. DOI: 10.1016/j.molcel.2019.04.025

[39] Zhao BS, Roundtree IA, He C. Post-transcriptional gene regulation by mRNA modifications. Nature Reviews Molecular Cell Biology. 2017;**18**(1):31–42. DOI: 10.1038/nrm.2016.132

[40] Yen YP, Lung TH, Liau ES, Wu CC, Huang GL, Hsu FY, Chang M, Yang ZD, Huang CY, Zheng Z, Zhao W, Hung JH, He C, Nie Q, Chen JA. The motor neuron m6A repertoire governs neuronal homeostasis and FTO inhibition mitigates ALS symptom manifestation. Nature Communications. 2025;**16**(1):4063. DOI: 10.1038/s41467-025-59117-2

[41] Yan C, Yao Y, Zhang Z, Li F, Fan D, Liu W, Fan X, Xu L, Liu Y, Wang S, Hu M, Yang Y, Tang Z. Gut microbiota-mediated betaine regulates skeletal muscle fiber type transition by affecting m(6)A RNA methylation and Myh7 expression. Gut Microbes. 2025;**17**(1):2545434. DOI: 10.1080/19490976.2025.2545434

[42] Deng K, Fan Y, Liang Y, Cai Y, Zhang G, Deng M, Wang Z, Lu J, Shi J, Wang F, Zhang Y. FTO-mediated demethylation of GADD45B promotes myogenesis through the activation of p38 MAPK pathway. Molecular Therapy - Nucleic Acids. 2021;**26**:34–48. DOI: 10.1016/j.omtn.2021.06.013

[43] Wu Z, Lu M, Liu D, Shi Y, Ren J, Wang S, Jing Y, Zhang S, Zhao Q, Li H, Yu Z, Liu Z, Bi S, Wei T, Yang YG, Xiao J, Belmonte JCI, Qu J, Zhang W, Ci W, Liu GH. A epitranscriptomic regulation of tissue homeostasis during primate aging. Nature Aging. 2023;**6**(3):705–721. DOI: 10.1038/s43587-023-00393-2

[44] Gilbert CJ, Rabolli CP, Golubeva VA, Sattler KM, Wang M, Ketabforoush A, Arnold WD, Lepper C, Accornero F. YTHDF2 governs muscle size through a targeted modulation of proteostasis. Nature Communications. 2024;**15**(1):2176. DOI: 10.1038/s41467-024-46546-8

[45] Palade GE. A small particulate component of the cytoplasm. The Journal of Biophysical and Biochemical Cytology. 1955;**1**(1):59–68. DOI: 10.1083/jcb.1.1.59

[46] Hoagland MB, Stephenson ML, Scott JF, Hecht LI, Zamecnik PC.A soluble ribonucleic acid intermediate in protein synthesis. The Journal of Biological Chemistry. 1958;**231**(1):241–257

[47] Lacatena RM, Cesareni G. Base pairing of RNA I with its complementary sequence in the primer precursor inhibits ColE1 replication. Nature. 1981;**294**(5842):623–626. DOI: 10.1038/294623a0

[48] Reyes AA, Marcum RD, He Y. Structure and Function of Chromatin

Remodelers. Journal of Molecular Biology. 2021;**433**(14):166929. DOI: 10.1016/j.jmb.2021.166929

[49] O'Brien J, Hayder H, Zayed Y, Peng C. Overview of microrna biogenesis, mechanisms of actions, and circulation. Frontiers in Endocrinology. 2018;**9**:402. DOI: 10.3389/fendo.2018.00402

[50] Li J, Wang L, Hua X, Tang H, Chen R, Yang T, Das S, Xiao J. CRISPR/Cas9-mediated miR-29b editing as a treatment of different types of muscle atrophy in mice. Molecular Therapy: the Journal of the American Society of Gene Therapy. 2020;**28**(5):1359–1372. DOI: 10.1016/j.ymthe.2020.03.005

[51] Qiu D, Zhang Y, Ni P, Wang Z, Yang L, Li F. Muscle-enriched microRNA-486-mediated regulation of muscular atrophy and exercise. Journal of Physiology and Biochemistry. 2024;**80**(4):795–809. DOI: 10.1007/s13105-024-01043-w

[52] Yang X, Xue P, Chen H, Yuan M, Kang Y, Duscher D, Machens HG, Chen Z. Denervation drives skeletal muscle atrophy and induces mitochondrial dysfunction, mitophagy and apoptosis via miR-142a-5p/MFN1 axis. Theranostics. 2020;**10**(3):1415–1432. DOI: 10.7150/thno.40857

[53] Li Y, Jiang J, Liu W, Wang H, Zhao L, Liu S, Li P, Zhang S, Sun C, Wu Y, Yu S, Li X, Zhang H, Qian H, Zhang D, Guo F, Zhai Q, Ding Q, Wang L, Ying H. microRNA-378 promotes autophagy and inhibits apoptosis in skeletal muscle. Proceedings of the National Academy of Sciences of the United States of America. 2018;**115**(46):E10849–e10858. DOI: 10.1073/pnas.1803377115

[54] Wang F, Wang J, He J, Li W, Li J, Chen S, Zhang P, Liu H, Chen X. Serum miRNAs miR-23a, 206, and 499 as potential biomarkers for skeletal muscle atrophy. BioMed Research International. 2017;**2017**:8361237. DOI: 10.1155/2017/8361237

[55] Qin M, Zhu J, Xing L, Fan Y, Luo J, Sun J, Chen T, Zhang Y, Xi Q. Adipose-derived exosomes ameliorate skeletal muscle atrophy via miR-146a-5p/IGF-1R signaling. Journal of Nanobiotechnology. 2024;**22**(1):754. DOI: 10.1186/s12951-024-02983-7

[56] Ma H, Jing Y, Zeng J, Ge J, Sun S, Cui R, Qian C, Qu S, Sheng H. Human umbilical cord mesenchymal stem cell-derived exosomes ameliorate muscle atrophy via the miR-132-3p/FoxO3 axis. Journal of Orthopaedic Translation. 2024;**49**:23–36. DOI: 10.1016/j.jot.2024.08.005

[57] Li Z, Liu C, Li S, Li T, Li Y, Wang N, Bao X, Xue P, Liu S. BMSC-derived exosomes inhibit dexamethasone-induced muscle atrophy via the miR-486-5p/FoxO1 axis. Frontiers in Endocrinology. 2021;**12**:681267. DOI: 10.3389/fendo.2021.681267

[58] Freire PP, Cury SS, Lopes LO, Fernandez GJ, Liu J, de Moraes LN, de Oliveira G, JS DO, de Moraes D, Cabral-Marques O, Dal-Pai-Silva M, Hu X, Wang DZ, Carvalho RF.Decreased miR-497-5p suppresses IL-6 induced atrophy in muscle cells. Cells. 2021;**10**(12). DOI: 10.3390/cells10123527

[59] Yu Y, Li X, Liu L, Chai J, Haijun Z, Chu W, Yin H, Ma L, Duan H, Xiao M. miR-628 promotes burn-induced skeletal muscle atrophy via targeting IRS1. International Journal of Biological

Sciences. 2016;**12**(10):1213–1224. DOI: 10.7150/ijbs.15496

[60] Li J, Yang T, Sha Z, Tang H, Hua X, Wang L, Wang Z, Gao Z, Sluijter JPG, Rowe GC, Das S, Yang L, Xiao J. Angiotensin II-induced muscle atrophy via PPARγ suppression is mediated by miR-29b. Molecular Therapy - Nucleic Acids. 2021;**23**:743–756. DOI: 10.1016/j.omtn.2020.12.015

[61] Murata M, Nonaka H, Komatsu S, Goto M, Morozumi M, Yamada S, Lin IC, Yamashita S, Tachibana H. Delphinidin prevents muscle atrophy and upregulates miR-23a expression. Journal of Agricultural and Food Chemistry. 2017;**65**(1):45–50. DOI: 10.1021/acs.jafc.6b03661

[62] Geng H, Song Q, Cheng Y, Li H, Yang R, Liu S, Hao L.MicroRNA 322 aggravates dexamethasone-induced muscle atrophy by targeting IGF1R and INSR. International Journal of Molecular Sciences. 2020;**21**(3). DOI: 10.3390/ijms21031111

[63] Todaka H, Higuchi T, Yagyu K, Sugiyama Y, Yamaguchi F, Morisawa K, Ono M, Fukushima A, Tsuda M, Taniguchi T, Sakamoto S. Overexpression of NF90-NF45 represses myogenic microRNA biogenesis, resulting in development of skeletal muscle atrophy and centronuclear muscle fibers. Molecular and Cellular Biology. 2015;**35**(13):2295–2308. DOI: 10.1128/mcb.01297-14

[64] Liu C, Wang M, Chen M, Zhang K, Gu L, Li Q, Yu Z, Li N, Meng Q. miR-18a induces myotubes atrophy by down-regulating IgfI. The International Journal of Biochemistry & Cell Biology. 2017;**90**:145–154. DOI: 10.1016/j.biocel.2017.07.020

[65] Pegoraro V, Marozzo R, Angelini C. MicroRNAs and HDAC4 protein expression in the skeletal muscle of ALS patients. Clinical Neuropathology. 2020;**39**(3):105–114. DOI: 10.5414/np301233

[66] Zhang Y, Wang T, Wang Z, Shi X, Jin J. Functions and therapeutic potentials of long noncoding RNA in skeletal muscle atrophy and dystrophy. Journal of Cachexia, Sarcopenia and Muscle. 2025;**16**(2):e13747. DOI: 10.1002/jcsm.13747

[67] Hitachi K, Honda M, Tsuchida K. The functional role of long non-coding RNA in myogenesis and skeletal muscle atrophy. Cells. 2022;**11**(15). DOI: 10.3390/cells11152291

[68] Yao Y, Yan C, Huang H, Wang S, Li J, Chen Y, Qu X, Bao Q, Xu L, Zhang Y, Fan D, He X, Liu Y, Zhang Y, Yang Y, Tang Z. LncRNA-MEG3 regulates muscle mass and metabolic homeostasis by facilitating SUZ12 Liquid-liquid phase separation. Advanced Science (Weinheim, Baden-Wurttemberg, Germany). 2025;**12**(23): e2417715. DOI: 10.1002/advs.202417715

[69] Cai B, Ma M, Yuan R, Zhou Z, Zhang J, Kong S, Lin D, Lian L, Li J, Zhang X, Nie Q. MYH1G-AS is a chromatin-associated lncRNA that regulates skeletal muscle development in chicken. Cellular & Molecular Biology Letters. 2024;**29**(1):9. DOI: 10.1186/s11658-023-00525-x

[70] Lv W, Jiang W, Luo H, Tong Q, Niu X, Liu X, Miao Y, Wang J, Guo Y, Li J, Zhan X, Hou Y, Peng Y, Wang J, Zhao S, Xu Z, Zuo B. Long noncoding RNA lncMREF promotes myogenic differentiation and muscle regeneration by interacting with the Smarca5/p300 complex. Nucleic Acids Research.

2022;**50**(18):10733–10755. DOI: 10.1093/nar/gkac854

[71] Hitachi K, Nakatani M, Tsuchida K. Long non-coding RNA myoparr regulates GDF5 expression in denervated mouse skeletal muscle. Non-coding RNA. 2019;**5**(2). DOI: 10.3390/ncrna5020033

[72] Zhang H, Wang F, Pang X, Zhou Y, Li S, Li W, Zhang P, Chen X. Decreased expression of H19/miR-675 ameliorates muscle atrophy by regulating the IGF1R/Akt/FoxO signaling pathway. Molecular Medicine (Cambridge, Mass). 2023;**29**(1):78. DOI: 10.1186/s10020-023-00683-w

[73] Qi T, Qin H, Yu F, Zhou Z, Chen Y, Liu P, Zeng H, Weng J. XLOC_015548 mitigates skeletal muscle atrophy via the Gadd45g/MEK/ERK pathway and redox regulation. Frontiers in Bioscience (Landmark Edition). 2025;**30**(4):36233. DOI: 10.31083/fbl36233

[74] Huang R, Tang X, Liu S, Sun L. Decoding CKD-induced muscle atrophy through the critical role of lncRNA GAS5 and pyroptosis. Molecular Therapy - Nucleic Acids. 2025;**36**(1):102451. DOI: 10.1016/j.omtn.2025.102451

[75] Zhang ZK, Li J, Guan D, Liang C, Zhuo Z, Liu J, Lu A, Zhang G, BT Z. Long noncoding RNA lncMUMA reverses established skeletal muscle atrophy following mechanical unloading. Molecular Therapy: the Journal of the American Society of Gene Therapy. 2018;**26**(11):2669–2680. DOI: 10.1016/j.ymthe.2018.09.014

[76] Jin JJ, Lv W, Xia P, Xu ZY, Zheng AD, Xj W, SS W, Zeng R, Luo HM, Li GL, Zuo B. Long noncoding RNA SYISL regulates myogenesis by interacting with polycomb repressive complex 2. Proceedings of the National Academy of Sciences of the United States of America. 2018;**115**(42):E9802–e9811. DOI: 10.1073/pnas.1801471115

[77] Zhu X, Yang T, Zheng Y, Nie Q, Chen J, Li Q, Ren X, Yin X, Wang S, Yan Y, Liu Z, Wu M, Lu D, Yu Y, Chen L, Chatterjee E, Li G, Cretoiu D, Bowen TS, Li J, Xiao J. EIF4A3-induced circular RNA circDdb1 promotes muscle atrophy through encoding a novel protein circDdb1-867aa. Advanced Science (Weinheim, Baden-Wurttemberg, Germany). 2024;**11**(45):e2406986. DOI: 10.1002/advs.202406986

[78] Chen R, Yang T, Jin B, Xu W, Yan Y, Wood N, Lehmann HI, Wang S, Zhu X, Yuan W, Chen H, Liu Z, Li G, Bowen TS, Li J, Xiao J. CircTmeff1 promotes muscle atrophy by interacting with TDP-43 and encoding a novel TMEFF1-339aa protein. Advanced Science (Weinheim, Baden-Wurttemberg, Germany). 2023;**10**(17):e2206732. DOI: 10.1002/advs.202206732

[79] Ma W, Wang C, Xiao J. A testlet diagnostic classification model with attribute hierarchies. Applied Psychological Measurement. 2023;**47**(3):183–199. DOI: 10.1177/01466216231165315

[80] Zhao X, Tang S, Lei Z, Shen X, Zhang Y, Han S, Yin H, Cui C. circAGO3 facilitates NF-κB pathway-mediated inflammatory atrophy in chicken skeletal muscle via the miR-34b-5p/TRAF3 axis. International Journal of Biological Macromolecules. 2024;**283**(3):137614. DOI: 10.1016/j.ijbiomac.2024.137614

[81] Zhao J, Zhao X, Shen X, Zhang Y, Zhang Y, Ye L, Li D, Zhu Q, Yin H. CircCCDC91 regulates chicken skeletal muscle development by sponging miR-15 family via activating IGF1-PI3K/AKT signaling pathway. Poultry Science. 2022;**101**(5):101803. DOI: 10.1016/j.psj.2022.101803

[82] Reinoso-Sánchez JF, Baroli G, Duranti G, Scaricamazza S, Sabatini S, Valle C, Morlando M, Casero Jr RA, Bozzoni I, Mariottini P, Ceci R, Cervelli M.Emerging role for linear and circular spermine oxidase RNAs in skeletal muscle physiopathology. International Journal of Molecular Sciences. 2020;**21**(21). DOI: 10.3390/ijms21218227

[83] Jacob F, Monod J. Genetic regulatory mechanisms in the synthesis of proteins. Journal of Molecular Biology. 1961;**3**:318–356. DOI: 10.1016/s0022-2836(61)80072-7

[84] Polack FP, Thomas SJ, Kitchin N, Absalon J, Gurtman A, Lockhart S, Perez JL, Pérez Marc G, Moreira ED, Zerbini C, Bailey R, Swanson KA, Roychoudhury S, Koury K, Li P, Kalina WV, Cooper D, Frenck Jr RW, Hammitt LL, Türeci Ö, Nell H, Schaefer A, Ünal S, Tresnan DB, Mather S, Dormitzer PR, Şahin U, Jansen KU, Gruber WC. Safety and efficacy of the BNT162b2 mRNA COVID-19 vaccine. The New England Journal of Medicine. 2020;**383**(27):2603–2615. DOI: 10.1056/NEJMoa2034577

[85] Jalkanen P, Kolehmainen P, Häkkinen HK, Huttunen M, Tähtinen PA, Lundberg R, Maljanen S, Reinholm A, Tauriainen S, Pakkanen SH, Levonen I, Nousiainen A, Miller T, Välimaa H, Ivaska L, Pasternack A, Naves R, Ritvos O, Österlund P, Kuivanen S, Smura T, Hepojoki J, Vapalahti O, Lempainen J, Kakkola L, Kantele A, Julkunen I. COVID-19 mRNA vaccine induced antibody responses against three SARS-CoV-2 variants. Nature Communications. 2021;**12**(1):3991. DOI: 10.1038/s41467-021-24285-4

[86] Magadum A, Kaur K, Zangi L. mRNA-based protein replacement therapy for the heart. Molecular Therapy: The Journal of the American Society of Gene Therapy. 2019 (4):785–793. DOI: 10.1016/j.ymthe.2018.11.018

[87] Roberts TC, Langer R, Wood MJA. Advances in oligonucleotide drug delivery. Nature Reviews Drug Discovery. 2020;**19**(10):673–694. DOI: 10.1038/s41573-020-0075-7

[88] Winkle M, El-Daly SM, Fabbri M, Calin GA. Noncoding RNA therapeutics - challenges and potential solutions. Nature Reviews Drug Discovery. 2021;**20**(8):629–651. DOI: 10.1038/s41573-021-00219-z

[89] Fire A, Xu S, Montgomery MK, Kostas SA, Driver SE, Mello CC. Potent and specific genetic interference by double-stranded RNA in Caenorhabditis elegans. Nature. 1998;**391**(6669):806–811. DOI: 10.1038/35888

[90] Zamore PD. RNA interference: Big applause for silencing in Stockholm. Cell. 2006;**127**(6):1083–1086. DOI: 10.1016/j.cell.2006.12.001

[91] Bumcrot D, Manoharan M, Koteliansky V, Sah DW. RNAi therapeutics: A potential new class of pharmaceutical drugs. Nature Chemical Biology. 2006;**2**(12):711–719. DOI: 10.1038/nchembio839

[92] Allerson CR, Sioufi N, Jarres R, Prakash TP, Naik N, Berdeja A, Wanders L, Griffey RH, Swayze EE, Bhat B. Fully 2'-modified oligonucleotide duplexes with improved in vitro potency and stability compared to unmodified small interfering RNA. Journal of Medicinal Chemistry. 2005;**48**(4):901–904. DOI: 10.1021/jm049167j

[93] Choung S, YJ K, Kim S, Park HO, Choi YC. Chemical modification of siRNAs to improve serum stability without loss of efficacy. Biochemical and Biophysical Research Communications. 2006;**342**(3): 919–927. DOI: 10.1016/j.bbrc.2006.02.049

[94] Khvorova A, Watts JK. The chemical evolution of oligonucleotide therapies of clinical utility. Nature Biotechnology. 2017;**35**(3):238–248. DOI: 10.1038/nbt.3765

[95] Khan T, Weber H, DiMuzio J, Matter A, Dogdas B, Shah T, Thankappan A, Disa J, Jadhav V, Lubbers L, Sepp-Lorenzino L, Strapps WR, Tadin-Strapps M. Silencing Myostatin Using Cholesterol-conjugated siRNAs Induces Muscle Growth. Molecular Therapy - Nucleic Acids. 2016;**5**(8): e342. DOI: 10.1038/mtna.2016.55

[96] Engelbeen S, Pasteuning-Vuhman S, Boertje-van der Meulen J, Parmar R, Charisse K, Sepp-Lorenzino L, Manoharan M, Aartsma-Rus A, van Putten M. Efficient Downregulation of Alk4 in Skeletal Muscle After Systemic Treatment with Conjugated siRNAs in a Mouse Model for Duchenne Muscular Dystrophy. Nucleic Acid Therapeutics. 2023;**33**(1):26–34. DOI: 10.1089/nat.2022.0021

[97] Wang S, Xu X, Liu Y, Jin J, Zhu F, Bai W, Guo Y, Zhang J, Zuo H, Xu Z, Zuo B. RIP-Seq of EZH2 Identifies TCONS-00036665 as a Regulator of Myogenesis in Pigs. Frontiers in Cell and Developmental Biology. 2020;**8**:618617. DOI: 10.3389/fcell.2020.618617

[98] Sun L, Si M, Liu X, Choi JM, Wang Y, Thomas SS, Peng H, Hu Z. Long-noncoding RNA Atrolnc-1 promotes muscle wasting in mice with chronic kidney disease. Journal of Cachexia, Sarcopenia and Muscle. 2018;**9**(5):962–974. DOI: 10.1002/jcsm.12321

[99] Monia BP, Lesnik EA, Gonzalez C, Lima WF, McGee D, Guinosso CJ, Kawasaki AM, Cook PD, Freier SM. Evaluation of 2'-modified oligonucleotides containing 2'-deoxy gaps as antisense inhibitors of gene expression. The Journal of Biological Chemistry. 1993;**268**(19):14514–14522

[100] Havens MA, Hastings ML. Splice-switching antisense oligonucleotides as therapeutic drugs. Nucleic Acids Research. 2016;**44**(14):6549–6563. DOI: 10.1093/nar/gkw533

[101] Lefebvre S, Bürglen L, Reboullet S, Clermont O, Burlet P, Viollet L, Benichou B, Cruaud C, Millasseau P, Zeviani M, et al. Identification and characterization of a spinal muscular atrophy-determining gene. Cell. 1995;**80**(1):155–165. DOI: 10.1016/0092-8674(95)90460-3.

[102] Wirth B. An update of the mutation spectrum of the survival motor neuron gene (SMN1) in autosomal recessive spinal muscular atrophy (SMA). Human Mutation. 2000;**15**(3):228–237. DOI: 10.1002/(SICI)1098-1004(200003)15:3<228::AID-HUMU3>3.0.CO;2-9

[103] Lunn MR, Wang CH. Spinal muscular atrophy. Lancet (London, England). 2008;**371**(9630):2120–2133. DOI: 10.1016/s0140-6736(08)60921-6

[104] Singh NN, Lawler MN, Ottesen EW, Upreti D, Kaczynski JR, Singh RN. An intronic structure enabled by a long-distance interaction serves as a novel target for splicing correction in spinal muscular atrophy. Nucleic Acids Research. 2013;**41**(17):8144–8165. DOI: 10.1093/nar/gkt609

[105] Ottesen EW. ISS-N1 makes the First FDA-approved Drug for Spinal Muscular Atrophy. Translational Neuroscience. 2017;**8**:1–6. DOI: 10.1515/tnsci-2017-0001

[106] Krützfeldt J, Rajewsky N, Braich R, Rajeev KG, Tuschl T, Manoharan M, Stoffel M. Silencing of microRNAs in vivo with 'antagomirs'. Nature. 2005;**438**(7068):685–689. DOI: 10.1038/nature04303

[107] Ebert MS, Neilson JR, Sharp PA. MicroRNA sponges: Competitive inhibitors of small RNAs in mammalian cells. Nature Methods. 2007;**4**(9): 721–726. DOI: 10.1038/nmeth1079

[108] Li J, Chan MC, Yu Y, Bei Y, Chen P, Zhou Q, Cheng L, Chen L, Ziegler O, Rowe GC, Das S, Xiao J. miR-29b contributes to multiple types of muscle atrophy. Nature Communications. 2017;**8**:15201. DOI: 10.1038/ncomms15201

[109] Liu Q, Yuan W, Yan Y, Jin B, You M, Liu T, Gao M, Li J, Gokulnath P, Vulugundam G, Li G, Xu B, Xiao J. Identification of a novel small-molecule inhibitor of miR-29b attenuates muscle atrophy. Molecular Therapy - Nucleic Acids. 2023;**31**:527–540. DOI: 10.1016/j.omtn.2023.02.003

[110] Greene MA, Worley GA, Udoka ANS, Powell RR, Bruce T, Klotz JL, Bridges Jr WC, Duckett SK. Use of AgomiR and AntagomiR technologies to alter satellite cell proliferation in vitro, miRNA expression, and muscle fiber hypertrophy in intrauterine growth-restricted lambs. Frontiers in Molecular Biosciences. 2023;**10**:1286890. DOI: 10.3389/fmolb.2023.1286890

[111] He Q, Qiu J, Dai M, Fang Q, Sun X, Gong Y, Ding F, Sun H. MicroRNA-351 inhibits denervation-induced muscle atrophy by targeting TRAF6. Experimental and Therapeutic Medicine. 2016;**12**(6):4029–4034. DOI: 10.3892/etm.2016.3856

[112] Yao H, Qian J, Bian X-T, Guo L, Tang K-L, Tao X. miR-27b-3p reduces muscle fibrosis during chronic skeletal muscle injury by targeting TGF-βR1/Smad pathway. Journal of Orthopaedic Surgery and Research. 2024;**19**(1):329. DOI: 10.1186/s13018-024-04733-9

[113] van der Werf S, Bradley J, Wimmer E, Studier FW, Dunn JJ. Synthesis of infectious poliovirus RNA by purified T7 RNA polymerase. Proceedings of the National Academy of Sciences of the United States of America. 1986;**83**(8):2330–2334. DOI: 10.1073/pnas.83.8.2330

[114] Maruggi G, Mallett CP, Westerbeck JW, Chen T, Lofano G, Friedrich K, Qu L, Sun JT, McAuliffe J, Kanitkar A, Arrildt KT, Wang K-F, McBee I, McCoy D, Terry R, Rowles A, Abrahim MA, Ringenberg MA, Gains MJ, Spickler C, Xie X, Zou J, Shi P-Y, Dutt T, Henao-Tamayo M, Ragan I, Bowen RA, Johnson R, Nuti S, Luisi K, Ulmer JB, Steff A-M, Jalah R, Bertholet S, Stokes AH, Yu D. A

self-amplifying mRNA SARS-CoV-2 vaccine candidate induces safe and robust protective immunity in preclinical models. Molecular Therapy: the Journal of the American Society of Gene Therapy. 2022;**30**(5):1897–1912. DOI: 10.1016/j.ymthe.2022.01.001

[115] Wesselhoeft RA, Kowalski PS, Anderson DG. Engineering circular RNA for potent and stable translation in eukaryotic cells. Nature Communications. 2018;**9**(1):2629. DOI: 10.1038/s41467-018-05096-6

[116] Wolff JA, Malone RW, Williams P, Chong W, Acsadi G, Jani A, Felgner PL. Direct gene transfer into mouse muscle in vivo. Science (New York, NY). 1990;**247**(4949):1465–1468. DOI: 10.1126/science.1690918

[117] Sahin U, Karikó K, Türeci Ö. mRNA-based therapeutics--developing a new class of drugs. Nature Reviews Drug Discovery. 2014;**13**(10):759–780. DOI: 10.1038/nrd4278

[118] Ai X, Yan B, Witman N, Gong Y, Yang L, Tan Y, Chen Y, Liu M, Lu T, Luo R, Wang H, Chien KR, Wang W, Fu W. Transient secretion of VEGF protein from transplanted hiPSC-CMs enhances engraftment and improves rat heart function post MI. Molecular Therapy: the Journal of the American Society of Gene Therapy. 2023;**31**(1):211–229. DOI: 10.1016/j.ymthe.2022.08.012

[119] Carlsson L, Clarke JC, Yen C, Gregoire F, Albery T, Billger M, Egnell A-C, Gan L-M, Jennbacken K, Johansson E, Linhardt G, Martinsson S, Sadiq MW, Witman N, Wang Q-D, Chen C-H, Wang Y-P, Lin S, Ticho B, Hsieh PCH, Chien KR, Fritsche-Danielson R. Biocompatible, Purified VEGF-A mRNA Improves Cardiac Function after Intracardiac Injection 1 Week Post-myocardial Infarction in Swine. Molecular Therapy - Methods & Clinical Development. 2018;**9**:330–346. DOI: 10.1016/j.omtm.2018.04.003

[120] Dowling JJ, Weihl CC, Spencer MJ. Molecular and cellular basis of genetically inherited skeletal muscle disorders. Nature Reviews Molecular Cell Biology. 2021;**22**(11):713–732. DOI: 10.1038/s41580-021-00389-z

[121] Stadelmann C, Di Francescantonio S, Marg A, Müthel S, Spuler S, Escobar H. mRNA-mediated delivery of gene editing tools to human primary muscle stem cells. Molecular Therapy - Nucleic Acids. 2022;**28**:47–57. DOI: 10.1016/j.omtn.2022.02.016

[122] Chen LL, Yang L. Regulation of circRNA Biogenesis. RNA Biology. 2015;**12**(4):381–388. DOI: 10.1080/15476286.2015.1020271

[123] Shen X, Zhang X, Ru W, Huang Y, Lan X, Lei C, Chen H. circINSR promotes proliferation and reduces apoptosis of embryonic myoblasts by sponging miR-34a. Molecular Therapy - Nucleic Acids. 2020;**19**:986–999. DOI: 10.1016/j.omtn.2019.12.032

[124] Lu D, Chatterjee S, Xiao K, Riedel I, Huang C-K, Costa A, Cushman S, Neufeldt D, Rode L, Schmidt A, Juchem M, Leonardy J, Büchler G, Blume J, Gern O-L, Kalinke U, Wen Tan WL, Foo R, Vink A, van Laake LW, van der Meer P, Bär C, Thum T. A circular RNA derived from the insulin receptor locus protects against doxorubicin-induced cardiotoxicity. European Heart Journal. 2022;**43**(42):4496–4511. DOI: 10.1093/eurheartj/ehac337

[125] Neufeldt D, Schmidt A, Mohr E, Lu D, Chatterjee S, Fuchs M, Xiao K, Pan W, Cushman S, Jahn C, Juchem M, Hunkler HJ, Cipriano G, Jürgens B, Schmidt K, Groß S, Jung M, Hoepfner J, Weber N, Foo R, Pich A, Zweigerdt R, Kraft T, Thum T, Bär C. Circular RNA circZFPM2 regulates cardiomyocyte hypertrophy and survival. Basic Research in Cardiology. 2024;**119**(4):613–632. DOI: 10.1007/s00395-024-01048-y

[126] Chen R, Wang SK, Belk JA, Amaya L, Li Z, Cardenas A, Abe BT, Chen C-K, Wender PA, Chang HY. Engineering circular RNA for enhanced protein production. Nature Biotechnology. 2023;**41**(2):262–272. DOI: 10.1038/s41587-022-01393-0

[127] Lavenniah A, Luu TDA, Li YP, Lim TB, Jiang J, Ackers-Johnson M, Foo RS-Y. Engineered circular RNA sponges act as miRNA inhibitors to attenuate pressure overload-induced cardiac hypertrophy. Molecular Therapy: the Journal of the American Society of Gene Therapy. 2020;**28**(6):1506–1517. DOI: 10.1016/j.ymthe.2020.04.006

[128] Li J, Chen R, Zheng YJ, Yuan WL, Yang TT, Zhu XL, Yan YW, Jin B, Xu WR, Zhang ZR, Li GP, Gokulnath P, Lei ZY, Xiao JJ. Engineered circular RNA circmiR-29b attenuates muscle atrophy by sponging MiR-29b. Advances in Therapy. 2022;**5**(7):2200029. DOI: 10.1002/adtp.202200029

[129] Keam SJ. Vamorolone: First Approval. Drugs. 2024;**84**(1):111–117. DOI: 10.1007/s40265-023-01986-2

[130] Pupo A, Fernández A, Low SH, François A, Suárez-Amarán L, Samulski RJ. AAV vectors: The Rubik's cube of human gene therapy. Molecular Therapy: the Journal of the American Society of Gene Therapy. 2022;**30**(12):3515–3541. DOI: 10.1016/j.ymthe.2022.09.015

[131] Falcucci L, Dooley CM, Adamoski D, Juan T, Martinez J, Georgieva AM, Mamchaoui K, Cirzi C, Stainier DYR. Transcriptional adaptation upregulates utrophin in Duchenne muscular dystrophy. Nature. 2025;**639**(8054):493–502. DOI: 10.1038/s41586-024-08539-x

[132] Baltusnikas J, Fokin A, Winkler J, Liobikas J. Long-term regulation of gene expression in muscle cells by systemically delivered siRNA. Journal of Controlled Release: Official Journal of the Controlled Release Society. 2017;**256**:101–113. DOI: 10.1016/j.jconrel.2017.04.037

[133] Lindley SR, Subbaiah KCV, Priyanka F, Poosala P, Ma Y, Jalinous L, West JA, Richardson WA, Thomas TN, Anderson DM. Ribozyme-activated mRNA trans-ligation enables large gene delivery to treat muscular dystrophies. Science (New York, NY). 2024;**386**(6723):762–767. DOI: 10.1126/science.adp8179

[134] Kenjo E, Hozumi H, Makita Y, Iwabuchi KA, Fujimoto N, Matsumoto S, Kimura M, Amano Y, Ifuku M, Naoe Y, Inukai N, Hotta A. Low immunogenicity of LNP allows repeated administrations of CRISPR-Cas9 mRNA into skeletal muscle in mice. Nature Communications. 2021;**12**(1):7101. DOI: 10.1038/s41467-021-26714-w

[135] Deleavey GF, Damha MJ. Designing chemically modified oligonucleotides for targeted gene silencing. Chemistry & Biology. 2012;**19**

(8):937–954. DOI: 10.1016/j.chembiol.2012.07.011

[136] Tasset A, Bellamkonda A, Wang W, Pyatnitskiy I, Ward D, Peppas N, Wang H. Overcoming barriers in non-viral gene delivery for neurological applications. Nanoscale. 2022;**14**(10):3698–3719. DOI: 10.1039/d1nr06939j

[137] Mitchell MJ, Billingsley MM, Haley RM, Wechsler ME, Peppas NA, Langer R. Engineering precision nanoparticles for drug delivery. Nature Reviews Drug Discovery. 2021;**20**(2):101–124. DOI: 10.1038/s41573-020-0090-8

[138] Bus T, Traeger A, Schubert US. The great escape: How cationic polyplexes overcome the endosomal barrier. Journal of Materials Chemistry B. 2018;**6**(43):6904–6918. DOI: 10.1039/c8tb00967h

[139] Rettig GR, Behlke MA. Progress toward in vivo use of siRNAs-II. Molecular Therapy: the Journal of the American Society of Gene Therapy. 2012;**20**(3):483–512. DOI: 10.1038/mt.2011.263

[140] Hald Albertsen C, Kulkarni JA, Witzigmann D, Lind M, Petersson K, Simonsen JB. The role of lipid components in lipid nanoparticles for vaccines and gene therapy. Advanced Drug Delivery Reviews. 2022;**188**:114416. DOI: 10.1016/j.addr.2022.114416.
114416

[141] Eygeris Y, Gupta M, Kim J, Sahay G. Chemistry of Lipid Nanoparticles for RNA Delivery. Accounts of Chemical Research. 2022;**55**(1):2–12. DOI: 10.1021/acs.accounts.1c00544

[142] Qu L, Yi Z, Shen Y, Lin L, Chen F, Xu Y, Wu Z, Tang H, Zhang X, Tian F, Wang C, Xiao X, Dong X, Guo L, Lu S, Yang C, Tang C, Yang Y, Yu W, Wang J, Zhou Y, Huang Q, Yisimayi A, Liu S, Huang W, Cao Y, Wang Y, Zhou Z, Peng X, Wang J, Xie XS, Wei W. Circular RNA vaccines against SARS-CoV-2 and emerging variants. Cell. 2022;**185**(10):1728–1744. DOI: 10.1016/j.cell.2022.03.044

[143] Ralvenius WT, Andresen JL, Huston MM, Penney J, Bonner JM, Fenton OS, Langer R, Tsai LH. Nanoparticle-Mediated Delivery of Anti-PU.1 siRNA via Localized Intracisternal Administration Reduces Neuroinflammation. Advanced Materials (Deerfield Beach, Fla). 2024;**36**(8):e2309225. DOI: 10.1002/adma.202309225

[144] Suo J, Li L, Tan W, Yin X, Wang J, Shao R, Sun S, Guo S-K, Feng J, Gao B-Q, Wang Y, Wei M-Y, Wang L, Feng H, Gao X, Hu P, Zheng X, Chen -L-L, Lei G, Huang Y, Zou W. Circular RNA-based protein replacement therapy mitigates osteoarthritis in male mice. Nature Communications. 2025;**16**(1):8480. DOI: 10.1038/s41467-025-63343-z

[145] Chen J, Xu Y, Zhou M, Xu S, Varley AJ, Golubovic A, Lu RXZ, Wang KC, Yeganeh M, Vosoughi D, Li B. Combinatorial design of ionizable lipid nanoparticles for muscle-selective mRNA delivery with minimized off-target effects. Proceedings of the National Academy of Sciences of the United States of America. 2023;**120**(50):e2309472120. DOI: 10.1073/pnas.2309472120

[146] Lin Y, Li M, Luo Z, Meng Y, Zong Y, Ren H, Yu X, Tan X, Liu F, Wei T, Cheng Q. Tissue-specific mRNA

delivery and prime editing with peptide–ionizable lipid nanoparticles. Nature Materials. 2025. DOI: 10.1038/s41563-025-02320-9

[147] Wang Y, Ma Z, Jiang L, Bojan N, Sha Y, Huang B, Ming L, Shen J, Pang W. Specific muscle targeted delivery of miR-130a loaded lipid nanoparticles: A novel approach to inhibit lipid accumulation in skeletal muscle and obesity. Journal of Nanobiotechnology. 2025;**23**(1):159. DOI: 10.1186/s12951-025-03225-0

[148] Takahashi Y, Takakura Y. Extracellular vesicle-based therapeutics: Extracellular vesicles as therapeutic targets and agents. Pharmacology & Therapeutics. 2023;**242**:108352. DOI: 10.1016/j.pharmthera.2023.108352. 108352

[149] O'Brien K, Breyne K, Ughetto S, Laurent LC, Breakefield XO. RNA delivery by extracellular vesicles in mammalian cells and its applications. Nature Reviews Molecular Cell Biology. 2020;**21**(10):585–606. DOI: 10.1038/s41580-020-0251-y

[150] Herrmann IK, Wood MJA, Fuhrmann G. Extracellular vesicles as a next-generation drug delivery platform. Nature Nanotechnology. 2021;**16**(7):748–759. DOI: 10.1038/s41565-021-00931-2

[151] Vader P, Mol EA, Pasterkamp G, Schiffelers RM. Extracellular vesicles for drug delivery. Advanced Drug Delivery Reviews. 2016;**106**:148–156. DOI: 10.1016/j.addr.2016.02.006

[152] Gupta D, Zickler AM, El Andaloussi S. Dosing extracellular vesicles. Advanced Drug Delivery Reviews. 2021;**178**:113961. DOI: 10.1016/j.addr.2021.113961. 113961

[153] Alvarez-Erviti L, Seow Y, Yin H, Betts C, Lakhal S, Wood MJA. Delivery of siRNA to the mouse brain by systemic injection of targeted exosomes. Nature Biotechnology. 2011;**29**(4):341–345. DOI: 10.1038/nbt.1807

[154] Gao X, Ran N, Dong X, Zuo B, Yang R, Zhou Q, Moulton HM, Seow Y, Yin H.Anchor peptide captures, targets, and loads exosomes of diverse origins for diagnostics and therapy. Science Translational Medicine. 2018;**10**(444). DOI: 10.1126/scitranslmed.aat0195

[155] Chen R, Yuan W, Zheng Y, Zhu X, Jin B, Yang T, Yan Y, Xu W, Chen H, Gao J, Li G, Gokulnath P, Vulugundam G, Li J, Xiao J. Delivery of engineered extracellular vesicles with miR-29b editing system for muscle atrophy therapy. Journal of Nanobiotechnology. 2022;**20**(1):304. DOI: 10.1186/s12951-022-01508-4

[156] Yang -M-M, Luo J-Y, Cai J, Qian J, Zhu X-Y, Zou H-Z, Li G-M, Dai J, Zhu H, Li W, Zhang L-D, Kang X. miRNA-125b-5p-rich exosomes derived from fibro-adipogenic progenitors promoting muscle regeneration through inducing pro-regenerative macrophages. Stem Cell Research & Therapy. 2025;**16**(1):363. DOI: 10.1186/s13287-025-04452-w

[157] Lin H, Yin L, Liu W, Li R, Jiang T, Yang M, Cao Y, Wang S, Yu Y, Chen C, Guo X, Wang W, Liu H, Dai Y, Yan J, Lin Y, Ding Y, Ruan C, Yang L, Wu T, Tao J, Chen L. Muscle-Derived Small Extracellular Vesicles Mediate Exercise-Induced Cognitive Protection in Chronic Cerebral Hypoperfusion. Advanced Science (Weinheim, Baden-Wurttemberg, Germany). 2025;**12**(27):e2410209. DOI: 10.1002/advs.202410209

[158] Komaki H, Takeshima Y, Matsumura T, Ozasa S, Funato M, Takeshita E, Iwata Y, Yajima H, Egawa Y, Toramoto T, Tajima M, Takeda S. Viltolarsen in Japanese Duchenne muscular dystrophy patients: A phase 1/2 study. Annals of Clinical and Translational Neurology. 2020;**7** (12):2393–2408. DOI: 10.1002/acn3.51235

[159] Vandenborne K, Walter GA, Straub V, Willcocks RJ, Forbes SC, Mercuri EM, Muntoni F, Ding K, Ennamuri S, Reid C, Murphy AP, Manfrini M, Mendell JR, Elkins JS, Rodino-Klapac LR. Quantitative Muscle Magnetic Resonance Outcomes in Patients With Duchenne Muscular Dystrophy: An Exploratory Analysis From the EMBARK Randomized Clinical Trial. JAMA Neurology. 2025;**82** (7):734–744. DOI: 10.1001/jamaneurol.2025.0992

Section 3

Bioactive Modulators of Skeletal Muscle Health

Chapter 6

Herbal Bioactives and Their Role in Modulating Inflammation and Oxidative Stress in Skeletal Muscle

Ahmed Karim, Saliha Mharchi, Mohamed Marghich and Abdelhay Addous

Abstract

Oxidative stress and inflammation play crucial roles in skeletal muscle atrophy and injury, impacting protein synthesis, proteolysis, and muscle regeneration. While inflammation is essential for muscle healing, excessive inflammatory responses can disrupt recovery. Various herbal bioactives have shown promise in modulating these processes. *Camellia sinensis*, *Curcuma longa*, *Panax ginseng,* and several other plants exhibit anti-inflammatory effects by reducing reactive oxygen species production, inhibiting cyclooxygenase activity, and downregulating inflammatory mediators such as IL-1, IL-6, and TNF-α. These plant-derived compounds can potentially optimize muscle healing and prevent atrophy. However, the use of anti-inflammatory drugs may have both beneficial and detrimental effects on muscle adaptations to exercise and training. Understanding the complex interplay between inflammation, oxidative stress, and muscle physiology is crucial for developing effective treatments for muscle-related disorders.

Keywords: phytotherapy, herbal bioactives, skeletal muscle, inflammation, oxidative stress, anti-inflammatory effect, antioxidant effect

1. Introduction

Skeletal muscle is one of the largest tissues in the human body. It is a collection of highly organized myofibers (muscle cells), which comprises about 45% of total body mass [1]. Beyond its role as the main protein reservoir, skeletal muscle not only controls locomotion but is also fundamental for vital functions such as breathing, eating, energy expenditure, as well as for glucose, amino acid, and lipid homeostasis, and for maintaining a high quality of life [2]. However, skeletal muscle is frequently affected by increasing age and other morbid conditions. These processes are triggered by various stimuli, including intense exercise, trauma, metabolic disorders, and aging.

Beyond its mechanical functions, skeletal muscle is also recognized as an endocrine organ capable of secreting bioactive molecules, such as myokines (IL-6, IGF-1, FGF2, etc.), synthesized and secreted by myocytes in response to muscle contraction, which influence systemic physiology [3]. These myokines play crucial roles in regulating muscle metabolism, hypertrophy, angiogenesis, and inflammatory processes [4]. Given its central role in both movement and metabolism, skeletal muscle is highly vulnerable to stressors that compromise its structural and functional integrity. Among the most critical stressors are inflammation and oxidative stress, two interrelated processes that profoundly impact muscle physiology [5].

Inflammation in skeletal muscle is a double-edged sword. On the one hand, acute inflammation following mechanical injury or exercise is essential for clearing damaged tissue and initiating the regeneration process [6]. This beneficial response is mediated by immune cell infiltration, the release of cytokines, and the activation of signaling cascades that stimulate repair [7]. However, when inflammation becomes chronic or excessive, the persistent production of cytokines is often accompanied by sustained oxidative stress, creating a harmful environment that accelerates tissue degeneration, promotes fibrosis, and impairs regeneration. This pathological state reduces muscle function and contributes to wasting conditions and metabolic disorders such as sarcopenia, cachexia, and metabolic syndrome [8–11].

Similarly, oxidative stress plays a paradoxical role in skeletal muscle physiology. At moderate levels, reactive oxygen species (ROS) act as signaling molecules that regulate adaptation to exercise by promoting mitochondrial biogenesis, angiogenesis, and antioxidant defense [12]. However, excessive ROS production or impaired antioxidant defenses can cause lipid peroxidation, protein oxidation, mitochondrial dysfunction, and contractile impairment, thereby exacerbating muscle weakness and fatigue [13]. Notably, oxidative stress and inflammation are tightly interconnected, as ROS activate redox-sensitive transcription factors, including NF-κB and AP-1, which, in turn, upregulate pro-inflammatory cytokines, creating a vicious cycle.

Given the adverse consequences of uncontrolled inflammation and oxidative stress, there is increasing interest in natural interventions capable of modulating these processes without the side effects often associated with synthetic drugs.

In this context, herbal bioactives, defined as naturally occurring plant-derived compounds with biological activity, have gained attention for their ability to modulate inflammatory and oxidative pathways. These compounds, including diverse classes of herbal bioactives such as polyphenols, flavonoids, terpenoids, saponins, and alkaloids, exhibit pleiotropic anti-inflammatory and antioxidant activities, making them promising candidates for skeletal muscle protection [14, 15]. For example, curcumin from *Curcuma longa* [16], resveratrol from grapes, epigallocatechin gallate from green tea [17], and ginsenosides from *Panax ginseng* have all demonstrated protective effects on skeletal muscle through modulation of signaling pathways like NF-κB, MAPK, and Nrf2 [18].

Importantly, emerging evidence from experimental and clinical studies highlights the ability of these compounds not only to attenuate muscle inflammation and oxidative damage but also to enhance recovery, endurance, and muscle regeneration. These findings have significant implications for athletic performance, aging populations, and patients suffering from chronic inflammatory or metabolic diseases.

This chapter explores the mechanisms of oxidative stress and inflammation in skeletal muscle, reviews major classes of herbal bioactives, and evaluates their role in preventing or mitigating muscle dysfunction.

DOI: http://dx.doi.org/10.5772/intechopen.1014219

2. Inflammation and oxidative stress in skeletal muscle

2.1 Mechanisms of skeletal muscle inflammation

Inflammation in skeletal muscle is a natural response to injury, infection, or exercise-induced stress. Key mediators include cytokines such as TNF-α, IL-6, and IL-1β, which are secreted by muscle cells and infiltrating immune cells like macrophages and neutrophils [7]. The activation of transcription factors NF-κB and MAPKs plays a crucial role in regulating inflammatory responses and gene expression in skeletal muscle [19]. NF-κB induces pro-inflammatory genes, including cytokines and chemokines, and regulates immune cell functions [20, 21] (**Figure 1**).

In inflammatory myopathies, NF-κB activation in blood vessel endothelium attracts inflammatory cells, while joint activation of NF-κB p65 and NFAT5 in myofibers reflects characteristic tissue damage patterns [22]. Both NF-κB and MAPKs are stimulated by exercise, promoting improvements in fuel homeostasis and preventing muscle atrophy [19]. However, chronic activation of these pathways can contribute to various pathologies, such as diabetes and cachexia [19]. The interplay between NF-κB and NFAT transcription factors, mediated by cytokines, represents a double-edged sword in muscle disease, promoting both regeneration and inflammatory cell damage [22, 23] (**Figure 2**).

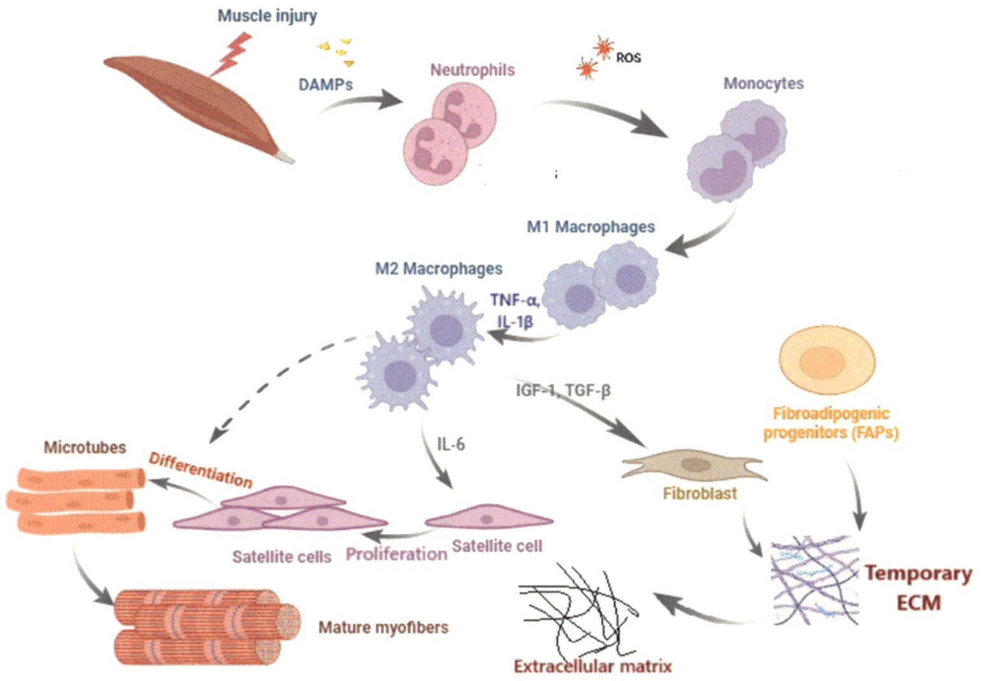

Figure 1.
Skeletal muscle repair following injury showing the three phases of muscle repair. Following muscle injury, inflammatory cells such as neutrophils, monocytes, and macrophages are recruited to the damaged tissue. In parallel, satellite cells become activated, proliferate, and differentiate into myotubes characterized by central nuclei, which subsequently mature into myofibers with peripheral nuclei. Meanwhile, fibroblasts and fibroadipogenic progenitors (FAPs) contribute to the production of a provisional extracellular matrix (ECM) that is later remodeled through the action of matrix metalloproteinases (MMPs) (Figure assembled by BioRender).

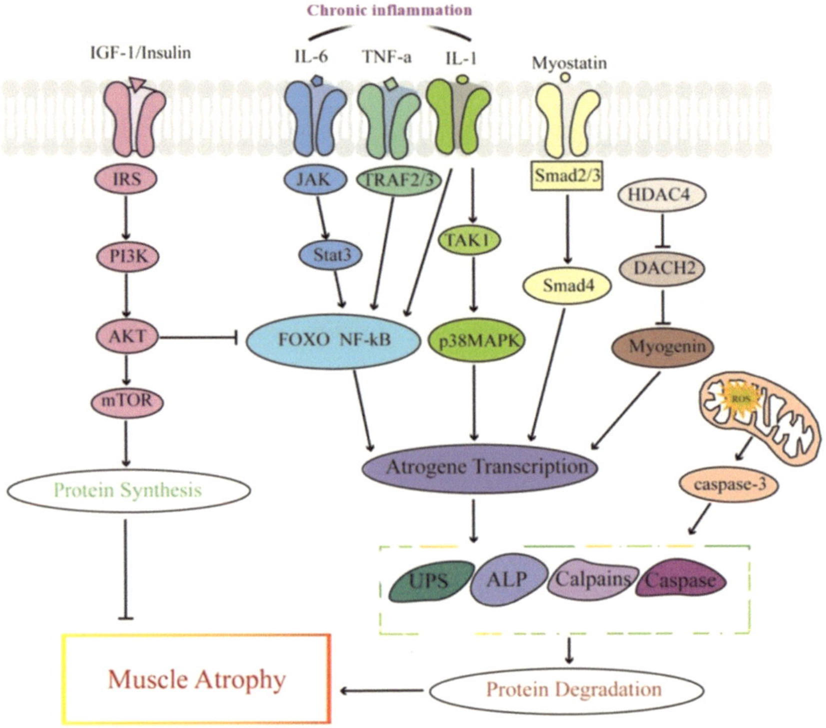

Figure 2.
Schematic representation of chronic inflammation skeletal muscle atrophy. IGF-1, insulin-like growth factor 1; IRS, insulin receptor substrate; PI3K, phosphoinositide 3-kinase; mTOR, mammalian target of rapamycin; IL-6, interleukin 6; JAK, Janus kinase; STAT3, signal transducer and activator of transcription 3; FOXO, forkhead transcription factor O-box subfamily; NF-κB, nuclear factor-κB; TNF-α, tumor necrosis factor-α; TRAF, TNF receptor-associated factor; IL-1, interleukin 1; TAK1, transforming growth factor β-activated kinase 1; HDAC4, histone deacetylase 4; DACH2, dachshund homolog 2; UPS, ubiquitin–proteasome system; ALP, autophagy–lysosome pathway; ROS, reactive oxygen species (referring to [24]).

2.2 Oxidative stress in skeletal muscle

Oxidative stress in skeletal muscle results from an imbalance between reactive oxygen species (ROS) production and antioxidant defense [25]. ROS are generated primarily in mitochondria during aerobic metabolism, but also through NADPH oxidases and xanthine oxidase pathways [26]. Excess ROS can damage proteins, lipids, and DNA, impairing contractile function and accelerating muscle fatigue. Chronic oxidative stress is implicated in sarcopenia, muscle atrophy, and metabolic disorders [12].

2.2.1 Sources of ROS in skeletal muscle

The primary source of ROS is the mitochondrial electron transport chain during aerobic metabolism, where electron leakage at complexes I and III results in the formation of superoxide anions (O_2• –) [26]. Beyond mitochondria, additional enzymatic systems contribute to ROS production, including:

- NADPH oxidases (NOX family), which generate superoxide at the sarcolemma and transverse tubules during muscle activation [27, 28].
- Xanthine oxidase, particularly during ischemia-reperfusion conditions [29].
- Phospholipase A2 and cyclooxygenase pathways, which can produce lipid-derived radicals during inflammation [30], are significant in the study of inflammatory processes.

2.2.2 Consequences of ROS overproduction

When ROS accumulate beyond the buffering capacity of antioxidants (such as superoxide dismutase, catalase, and glutathione peroxidase), they can damage multiple cellular components:

- Proteins: oxidation of contractile proteins (e.g., actin, myosin) and enzymes reduces force production and metabolic efficiency [2].
- Lipids: peroxidation of membrane phospholipids disrupts sarcolemmal integrity and calcium homeostasis [3].
- DNA: oxidative lesions in mitochondrial and nuclear DNA impair transcriptional regulation and mitochondrial biogenesis [6].

This molecular damage directly impairs excitation-contraction coupling, accelerates muscle fatigue, and contributes to the loss of muscle mass and function.

2.2.3 Oxidative stress and skeletal muscle pathophysiology

Chronic oxidative stress is implicated in several pathological states:

- Sarcopenia (age-related muscle decline), where lifelong accumulation of ROS damages mitochondria and reduces regenerative capacity (**Figure 3**) [31].

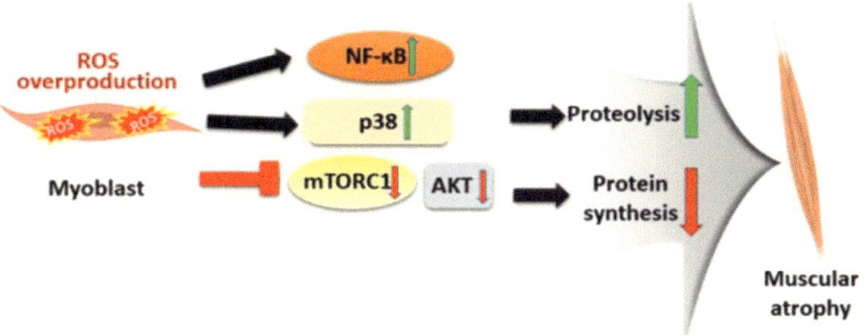

Figure 3.
Effects of reactive oxygen species (ROS) on muscle cells. During aging, skeletal muscle exhibits an increased production of reactive oxygen species (ROS). Elevated ROS levels inhibit the AKT/mTORC1 signaling pathway, thereby reducing protein synthesis and contributing to the development of muscle atrophy. In addition, excessive ROS activates NF-κB and p38 MAPK, which enhance proteolytic processes and further promote muscle wasting (referring to [32]).

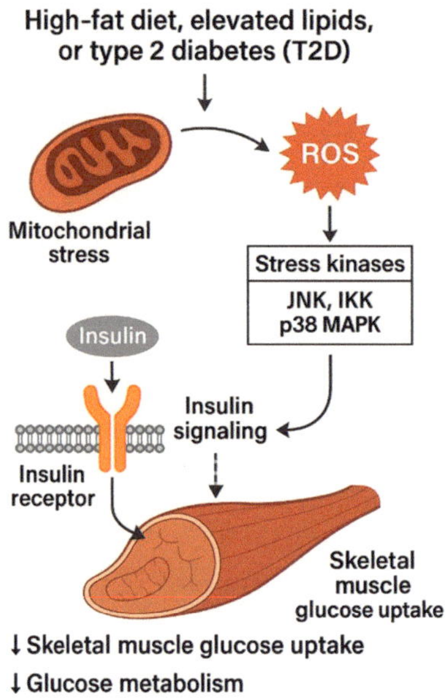

Figure 4.
Cellular origins and effects of ROS in skeletal muscle during insulin resistance. A high-fat diet, elevated circulating lipids, or type 2 diabetes (T2D) increases mitochondrial stress in skeletal muscle, leading to excessive production of reactive oxygen species (ROS). These ROS activate stress-responsive kinases such as JNK, IKK, and p38 MAPK, which interfere with the insulin signaling cascade. As a result, insulin receptor signaling becomes impaired, ultimately reducing skeletal muscle glucose transport and causing defects in intracellular glucose metabolism.

- Disuse atrophy and immobilization, conditions associated with decreased antioxidant enzyme expression [33].
- Metabolic disorders, such as insulin resistance and type 2 diabetes, where ROS alter insulin signaling pathways in skeletal muscle (**Figure 4**) [33].
- Inflammatory myopathies, in which ROS act synergistically with pro-inflammatory cytokines to exacerbate tissue injury [34].

2.3 Interplay between inflammation and oxidative stress

Inflammation and oxidative stress are intimately interconnected processes that often reinforce one another in skeletal muscle pathology. Reactive oxygen species (ROS) play a central role in this crosstalk. Excessive ROS production can activate redox-sensitive transcription factors, particularly nuclear factor kappa B (NF-κB) and activator protein-1 (AP-1), which upregulate the expression of pro-inflammatory cytokines such as TNF-α, IL-1β, and IL-6 [35]. This response amplifies the inflammatory cascade within muscle tissue.

On the other hand, inflammation itself stimulates ROS generation. Activated immune cells, including macrophages and neutrophils, release large amounts of ROS and reactive nitrogen species (RNS) through enzymatic systems such as NADPH oxidase, inducible nitric oxide synthase (iNOS), and myeloperoxidase [36]. These species contribute to oxidative modifications of proteins, lipid peroxidation, and mitochondrial dysfunction, further compromising muscle integrity [12].

This bidirectional interaction creates a self-perpetuating vicious cycle:

- ROS induce inflammation via NF-κB activation.
- Inflammation amplifies ROS production through immune cell activation.
- Together, they promote oxidative damage, impaired calcium handling, contractile dysfunction, and delayed muscle regeneration [31].

In chronic conditions, such as sarcopenia, muscular dystrophies, obesity, and type 2 diabetes, this persistent oxidative–inflammatory loop contributes to muscle atrophy, insulin resistance, and impaired recovery after injury [37].

Therefore, a deeper understanding of this interplay is crucial for identifying therapeutic strategies. Antioxidant compounds (e.g., polyphenols, vitamins C and E, N-acetylcysteine) and anti-inflammatory bioactives (e.g., curcumin, resveratrol, omega-3 fatty acids) have been shown to attenuate this cycle by modulating redox balance and suppressing pro-inflammatory signaling [38]. Such interventions highlight the potential of herbal bioactives in restoring muscle homeostasis and function.

3. Herbal bioactives: Classes and mechanisms of action

Phytotherapy, the use of medicinal plants and their bioactive compounds for health promotion and disease management, has gained significant attention in modern healthcare [39].

Plants produce a wide range of secondary metabolites, including polyphenols, alkaloids, terpenoids, and saponins, which have been associated with antioxidant, anti-inflammatory, and metabolic modulatory effects [14, 40] (**Table 1**). These properties make them promising agents for the management of disorders associated with oxidative stress and inflammation in skeletal muscle [14].

Herbal bioactives have been investigated for their potential to reduce muscle fatigue, enhance recovery, and mitigate degenerative processes resulting from chronic oxidative stress and inflammation [12, 41]. While conventional pharmacological approaches often provide symptomatic relief, they can be associated with adverse effects, highlighting the growing interest in complementary or alternative plant-based therapies [40].

The integration of medicinal plants into preventive and therapeutic strategies underscores their relevance in supporting muscle health and improving functional outcomes [38, 39].

3.1 Plant-derived secondary metabolites: Definition and classification

Plant-derived bioactive compounds predominantly consist of secondary metabolites, which are organic molecules not directly involved in the primary growth, development, or reproduction of plants. These compounds serve important ecological functions, such as protecting plants against pathogens, herbivores, and environmental stresses [42, 43].

	Bioactive compound	Class	Subfamily	Family	Botanical/ dietary origin	Part of plant	Antioxidant and anti-inflammatory effects on skeletal muscle	Main mechanisms of action in skeletal muscle	Effective concentration EC50	References
1	Curcumin	Polyphenol	Diaryl heptanoid	Curcuminoid	*Curcuma longa* (Turmeric)	Rhizome	Protects C2C12 cells from oxidativeinjury and inflammation.	Activates Nrf2; increases HO-1, NQO1, and GCLC; reduces ROS; inhibits NF-κB; maintains mitochondrial integrity.	1–30 μg/mL	[65, 66]
2	Resveratrol	Polyphenol	Stilbenoid	Stilbene	*Vitis vinifera* (Grapevine)	Grapes	Attenuates palmitate-induced inflammation in C2C12 myotubes; reduces macrophage infiltration and cytokines in HFD-induced obese mice.	Increases TNF-α and IL-6; reduces JNK, p38, ERK1/2, and IKKα/β phosphorylation; inhibits NF-κB; reduces ROS; improves insulin sensitivity (via AMPK activation).	C2C12: anti-inflammatory effects are typically observed at 10–50 μM.	[49, 67–69]
3	Epigallocatechin gallate	Polyphenol	Flavan-3-ol	Catechin	*Camellia sinensis* (Green tea)	Leaves	Preserves muscle mass and reduces atrophy in models of sarcopenia and disuse-induced atrophy.	Powerful ROS scavenger; decreases proteolysis through the ubiquitin–proteasome pathway; increases anabolic factors; modulates NF-κB and MAPK signaling.	Rat: oral doses around 50 mg/kg/day; in vitro effects are often observed at 5–25 μM.	[70, 71]

DOI: http://dx.doi.org/10.5772/intechopen.1014219

	Bioactive compound	Class	Subfamily	Family	Botanical/ dietary origin	Part of plant	Antioxidant and anti-inflammatory effects on skeletal muscle	Main mechanisms of action in skeletal muscle	Effective concentration EC50	References
4	Quercetin	Polyphenol	Flavonol	Flavonoid	*Capparis spinosa* (Capers)	Flower buds	Protects C2C12 cells from dexamethasone-induced atrophy; decreases oxidative stress and apoptosis; prevents unloading-induced muscle loss.	Modulation of Bax/Bcl-2 and $\Delta\Psi$m; reduction of ROS; decrease of MuRF1 and Atrogin-1; inhibition of NF-κB and pro-inflammatory cytokines (TNF-α, IL-6).	C2C12: protection observed between approximately 5–50 μM.	[32, 72–74]
5	Kampferol	Polyphenol	Flavonol	Flavonoid	*Brassica oleracea* (Kale), *Capparis spinosa* (Capers)	Leaves	Reduces LPS-induced muscle atrophy and improves exercise performance; restores the antioxidant system.	Inhibits KLF15, reduces atrogin-1 and MuRF1; increases SOD and catalase; enhances mitochondrial biogenesis through PI3K/AKT, MAPK, and mTOR pathways.	C2C12: typical effects observed at 5–20 μM.	[72, 75].

	Bioactive compound	Class	Subfamily	Family	Botanical/ dietary origin	Part of plant	Antioxidant and anti-inflammatory effects on skeletal muscle	Main mechanisms of action in skeletal muscle	Effective concentration EC50	References
6	Apigenin	Polyphenol	Flavone	Flavonoid	*Petroselinum crispum* (parsley), *Apium graveolens* (celery), *Matricaria chamomilla* (chamomile)	Leaves, flowers	Improves obesity-induced muscle atrophy; enhances mitochondrial biogenesis and exercise capacity; protects C2C12 cells from palmitate-induced atrophy.	Activates AMPK; increases mitochondrial function (OCR, complexes I–IV, ATP); reduces MuRF1 and Atrogin-1; decreases inflammatory cytokines.	C2C12: effects observed in the 5–25 μM range.	[76, 77]
7	Luteolin	Polyphenol	Flavone	Flavonoid	*Apium graveolens* (Celery), *Capsicum annuum* (Green pepper), *Thymus vulgaris* (Thyme)	Leaves, seeds, leaves	Protects against LPS- and obesity-induced muscle atrophy; strengthens antioxidant defenses in C2C12 cells.	Strongly activates Nrf2 and its target genes; reduces atrogin-1 in LPS-treated C2C12 cells; inhibits ferroptosis and prevents muscle mass loss under obesity or estrogen-deficient conditions.	C2C12: effects reported at approximately 5–20 μM.	[78, 79]

DOI: http://dx.doi.org/10.5772/intechopen.1014219

	Bioactive compound	Class	Subfamily	Family	Botanical/ dietary origin	Part of plant	Antioxidant and anti-inflammatory effects on skeletal muscle	Main mechanisms of action in skeletal muscle	Effective concentration EC50	References
8	Naringin/ Naringenin	Polyphenol	Flavanone	Flavonoid	*Citrus sp* (Orange) lemon, Grapefruit)	Fruits	Improves insulin sensitivity, prevents atrophy, and increases oxidative fibers in the skeletal muscle of obese rats and gestational diabetes models.	Active AMPK; ↑ GLUT4, PI3K, Akt; ↑ fibers slow-twitch et métabolisme aérobie; ↓ IL-6 et autres cytokines.	In vitro (C2C12): effects around 10–50 μM; in vivo: naringin ~ 50–100 mg/kg/day.	[80–82]
9	Cyanidin-3-O-glucoside	Polyphenol	Anthocyanin	Flavonoid	*Rubus fruticosus* (Berries), *Phaseolus vulgaris* (Black beans)	Fruits	Enhances insulin sensitivity and muscle metabolism; increases glucose uptake in C2C12 cells; improves HFD-induced insulin resistance.	Activates PPARs; increases GLUT4 and IRS-1; enhances fatty-acid oxidation; decreases TNF-α and IL-6; exerts nutrigenomic effects on metabolic pathways.	C2C12: effects reported at approximately 10–50 μM.	[83–85]
10	Mixed anthocyanins	Polyphenol	Anthocyanin	Flavonoid	*Vaccinium myrtillus* (Blueberry), *Ipomoea batatas* (Sweet potato)	Fruits	Reduces exercise-induced oxidative stress and inflammation; improves muscle recovery.	Powerful ROS scavenging; AMPK activation; increased GLUT4 expression; reduced TNF-α, IL-6, and CRP; improvement of DOMS markers and performance.	Human doses vary (≈ 200–500 mg/day of anthocyanins).	[86, 87]

	Bioactive compound	Class	Subfamily	Family	Botanical/ dietary origin	Part of plant	Antioxidant and anti-inflammatory effects on skeletal muscle	Main mechanisms of action in skeletal muscle	Effective concentration EC50	References
11	proanthocyanidins	Polyphenol	Condensed tannins	Flavonoids	*Vitis vinifera* (Grapevine)	Grape seeds	Increases muscle mass, oxidative enzymes, and fatigue resistance; decreases inflammation and oxidative stress.	Increases SIRT1, PGC-1α, and Nrf2; enhances mitochondrial biogenesis and Krebs cycle enzymes; modulates the microbiota–muscle axis and pro-inflammatory cytokines.	In vivo: 50–250 mg/kg/day.	[88]
12	Rutin	Polyphenol	Flavonol glycoside	Flavonoid	*Fagopyrum esculentum* (Buckwheat), *Ruta graveolens* (Rue)	Seeds, leaves	Prevents dexamethasone-induced muscle loss in C2C12 cells and mice; improves myotube diameter and fusion.	Controls FOXO3-dependent signaling; decreases the expression of atrogin-1 and MuRF1; strengthens the antioxidant system and reduces ROS.	0.1–5 μM	[89]
13	Berberine	Alkaloid	Isoquinoline alkaloid	Benzylisoquinoline	*Berberis spp.*, *Coptis chinensis* (Goldthread)	Roots	Reduces inflammation and oxidative stress; improves insulin sensitivity.	Reduces cytokines and ROS; enhances insulin pathways; protects mitochondria; inhibits NF-κB.	5-20 μM	[51, 90]
	Morphine	Alkaloid	Opiate	Alkaloid	*Papaver somniferum* (Opium poppy)		Anti-inflammatory effects on muscle damage	Modulates opioid receptors, reduces cytokines, and decreases oxidative stress.	1-10 μM	

DOI: http://dx.doi.org/10.5772/intechopen.1014219

	Bioactive compound	Class	Subfamily	Family	Botanical/ dietary origin	Part of plant	Antioxidant and anti-inflammatory effects on skeletal muscle	Main mechanisms of action in skeletal muscle	Effective concentration EC50	References
14	Chlorogenic acid	Phenolic acid	Hydroxycinnamic acid	Cinnamate derivative	*Coffea arabica* (Coffee)	Seeds	Enhances glucose uptake in insulin-resistant C2C12 cells; contributes to antifatigue effects in exercise models.	Increases GLUT4, insulin receptor, IRS-1, p-Akt, and p-AMPK; modulates redox and mitochondrial pathways; reduces ROS and oxidative stress.	10-100 μM;	[91]
15	Caffeic acid	Polyphenol	Phenolic acid	Hydroxycinnamic acid	*Coffea arabica* (Coffee)	Seeds	Strong antioxidant; reduces muscle inflammation	Reduces ROS, inhibits NF-κB, and enhances antioxidant enzymes.	80 mg/kg/day	[92]
16	Rosmarinic acid	Polyphenol	Phenolic acid	Hydroxycinnamic acid	*Rosmarinus officinalis* (Rosemary)	Leaves	Protects muscles from oxidative and inflammatory stress.	Neutralizes ROS, activates Nrf2, inhibits NF-κB and COX-2, and protects mitochondria.	1-20 μM	[93]
18	Limonene	Terpenoid	Monoterpene	Essential oil component	*Citrus aurantifolia*	Leaves	Reduces muscle oxidative stress and fatigue; provides anti-inflammatory protection in muscle fibers.	Reduces ROS; increases SOD and catalase; inhibits NF-κB and COX-2; protects mitochondrial function.	10-100 μM	[94]

	Bioactive compound	Class	Subfamily	Family	Botanical/ dietary origin	Part of plant	Antioxidant and anti-inflammatory effects on skeletal muscle	Main mechanisms of action in skeletal muscle	Effective concentration EC50	References
19	β-Carotene	Carotenoid	Tetraterpene	Carotenoid	*Daucus carota* (Carrots)	Roots	Protects myocytes against oxidative damage; decreases muscle inflammation.	Radical scavenging; activation of Nrf2; enhancement of antioxidant enzymes; reduction of TNF-α and IL-6 through NF-κB inhibition.	C2C12 cells: 20–40 μM	[95]
20	Lycopene	Carotenoid	Tetraterpene	Carotenoid	*Solanum lycopersicum* (Tomatoes)	skin and seeds	Strong antioxidant effect in myotubes; reduces muscle atrophy and inflammation.	Potent ROS scavenger; activates Nrf2/HO-1; suppresses NF-κB and iNOS; protects mitochondrial function.	5–15 μM	[96]

Table 1.
Summary of herbal bioactive molecules targeting skeletal muscle atrophy, oxidative stress, and inflammation.

Through phytotherapy, these secondary metabolites are employed for their capacity to modulate physiological processes, regulate oxidative balance, and support the prevention or management of various disorders [43]. Their structural diversity underlies a wide range of pharmacological activities, making them essential components in medicinal plant research, functional foods, and nutraceutical development.

Secondary metabolites are generally classified into several major chemical families according to their biosynthetic origin and chemical structure [42, 44]:

- Polyphenols are characterized by multiple phenolic groups. This class includes flavonoids (quercetin, catechins, anthocyanins), phenolic acids (caffeic acid, chlorogenic acid), and stilbenes (resveratrol). Polyphenols are abundant in fruits, vegetables, tea, coffee, and wine, and are widely studied for their health-promoting properties [43, 45] (**Figure 5**).

  Polyphenols have been linked to improved muscle recovery, enhanced mitochondrial biogenesis, and protection against oxidative damage in muscle tissue [46, 47]. For example, resveratrol, a stilbene found in grapes (*Vitis vinifera*), and epigallocatechin gallate (EGCG) from green tea (*Camellia sinensis*) have been associated with the attenuation of muscle inflammation and fatigue [48, 49].

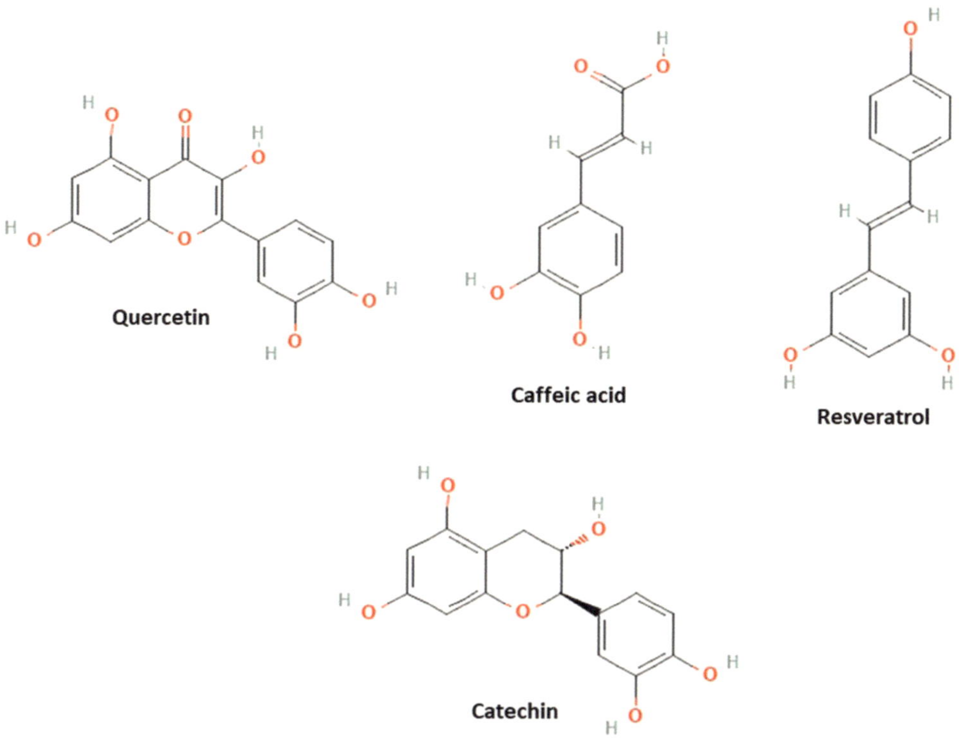

Figure 5.
Representative chemical structures of selected plant-derived polyphenols.

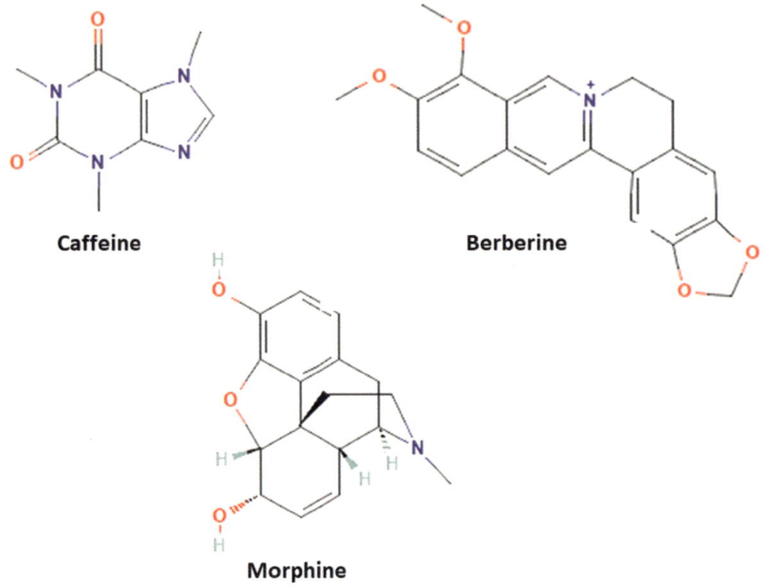

Figure 6.
Representative chemical structures of selected plant-derived alkaloids.

- Alkaloids: nitrogen-containing heterocyclic compounds such as caffeine (*Coffea arabica*), berberine (*Berberis vulgaris*), and morphine (*Papaver somniferum*). Alkaloids display diverse bioactivities, including metabolic and neuromodulatory effects. These compounds exhibit a wide spectrum of pharmacological properties, including antioxidant, anti-inflammatory, neuromodulatory, and anticancer effects [50] (**Figure 6, Table 1**). Some alkaloids, such as berberine, have been reported to modulate energy metabolism and protect against muscle atrophy [51].

- Terpenoids and saponins: Terpenoids represent the largest class of plant secondary metabolites, with diverse structural subgroups such as monoterpenes, diterpenes, triterpenes, carotenoids, and saponins. Notable examples include ginsenosides (*Panax ginseng*), the triterpenoid saponins, which are well-documented for their capacity to enhance physical performance, reduce exercise-induced fatigue, and improve skeletal muscle regeneration [18]; carotenoids such as β-carotene and lycopene; and essential oil constituents like limonene [42, 52] (**Figure 7, Table 1**). Similarly, ursolic acid, a pentacyclic triterpenoid found in apples and rosemary, has been studied for its anabolic effects on skeletal muscle [53].

The remarkable chemical diversity of these secondary metabolites underpins their contribution to skeletal muscle health by counteracting pathophysiological processes such as inflammation and oxidative stress. This diversity also explains their wide-ranging pharmacological potential, reinforcing their central importance in phytotherapy.

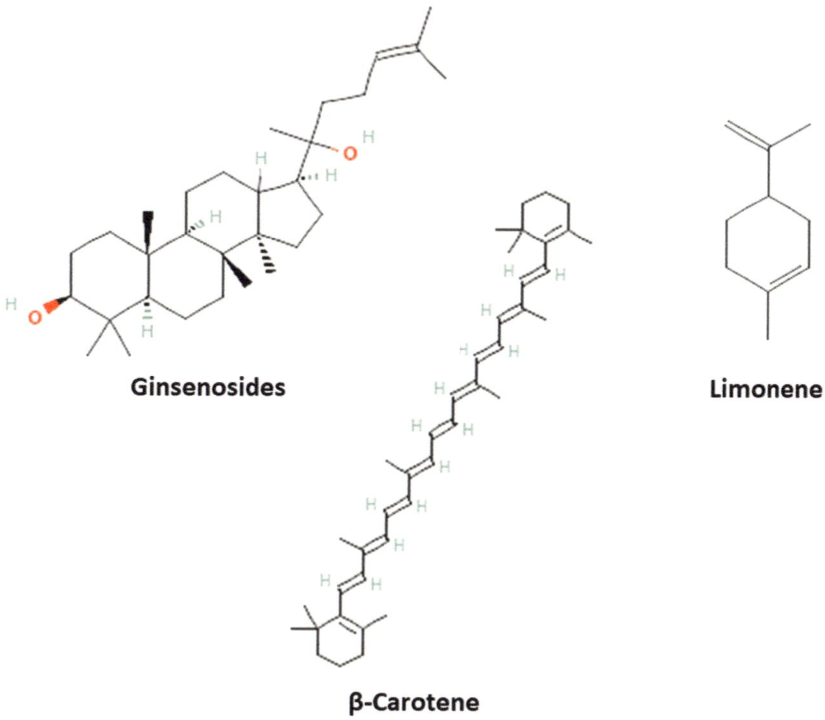

Figure 7.
Representative chemical structures of selected plant-derived terpenoids.

Owing to their multifunctional properties and structural versatility, these compounds provide a strong foundation for the development of herbal-based strategies aimed not only at managing muscle disorders but also at addressing a broader spectrum of health conditions linked to oxidative and inflammatory imbalances.

3.2 Mechanisms of action of herbal bioactives in muscles

3.2.1 Anti-inflammatory mechanisms

Chronic inflammation is a key contributor to skeletal muscle dysfunction, leading to impaired regeneration, increased proteolysis, and reduced contractile performance. Herbal bioactives exert potent anti-inflammatory effects by targeting major signaling pathways such as NF-κB and MAPKs, both of which regulate the transcription of pro-inflammatory cytokines.

Curcumin, a polyphenolic compound from *Curcuma longa*, inhibits the phosphorylation and degradation of IκBα, thereby preventing NF-κB p65 translocation into the nucleus and suppressing the transcription of cytokines such as TNF-α, IL-6, and IL-1β [54, 55]. Similarly, epigallocatechin gallate, the major catechin in green tea, attenuates inflammatory signaling by inhibiting p38 MAPK and JNK activation, ultimately reducing inflammatory gene expression in muscle cells [45]. Other compounds, such as quercetin and ginsenosides, also demonstrate strong anti-inflammatory activities by downregulating COX-2 and inducible nitric oxide synthase (iNOS) [56].

These effects collectively mitigate inflammation-induced muscle damage, enhance recovery following injury or exercise, and protect against chronic muscle pathologies such as sarcopenia and cachexia.

3.2.2 Antioxidant mechanisms

Oxidative stress, resulting from an imbalance between ROS production and antioxidant defenses, contributes to mitochondrial dysfunction, protein oxidation, and muscle fatigue. Herbal bioactives counteract oxidative stress through both direct and indirect mechanisms.

Directly, many polyphenols, such as flavonoids (quercetin, kampferol) and phenolic acids (rosmarinic acid), act as radical scavengers, neutralizing superoxide anions, hydroxyl radicals, and peroxynitrite [32, 57, 58]. Indirectly, they activate the Nrf2/ARE signaling pathway, leading to the upregulation of endogenous antioxidant enzymes, including superoxide dismutase (SOD), catalase (CAT), and glutathione peroxidase (GPx) [59]. For example, resveratrol enhances Nrf2 nuclear translocation, boosting cellular resilience against oxidative stress in skeletal muscle [60].

By maintaining redox homeostasis, these compounds reduce oxidative modifications of proteins and lipids, preserve mitochondrial integrity, and delay the onset of muscle fatigue, particularly in aging or high-intensity exercise contexts.

3.2.3 Effects on muscle metabolism and mitochondrial function

Mitochondria are central to muscle metabolism, energy production, and endurance. Dysfunctional mitochondria are strongly associated with muscle weakness, insulin resistance, and degenerative conditions. Herbal bioactives enhance mitochondrial function by modulating key regulators of energy metabolism.

Compounds such as ginsenosides (from Panax ginseng) and resveratrol stimulate AMPK (AMP-activated protein kinase) and PGC-1α (peroxisome proliferator-activated receptor γ co-activator 1-α), master regulators of mitochondrial biogenesis and oxidative phosphorylation [60, 61]. These activations increase mitochondrial number and efficiency, improve ATP production, and reduce ROS accumulation. Additionally, polyphenols improve fatty acid oxidation and glucose uptake, thereby enhancing metabolic flexibility in skeletal muscle [62].

These effects contribute not only to improved endurance and reduced fatigue during exercise but also to faster recovery after muscle injury. Furthermore, they hold therapeutic potential in conditions such as sarcopenia, muscular dystrophy, and metabolic disorders.

4. Experimental evidence

4.1 *In vitro* evaluation of antioxidant and anti-inflammatory effects in skeletal muscle cells

An established model for evaluating the bioactivity of plant-derived compounds involves the use of the murine myoblast cell line, C2C12 [63].

C2C12 murine myoblast cells were cultured in Dulbecco's modified eagle medium (DMEM), supplemented with 10% fetal bovine serum (FBS) and 1% penicillin–streptomycin, at 37 °C in a humidified atmosphere containing 5% CO_2.

For differentiation into myotubes, cells were seeded at an appropriate density and, upon reaching 70–80% confluence, the medium was replaced with DMEM containing 2% horse serum, which was refreshed every 48 hours until mature myotubes were formed (approximately 5–7 days).

To evaluate the protective role of curcumin, cells were pretreated with curcumin at different concentrations (5, 10, and 20 μM) for 2 hours prior to the induction of oxidative and inflammatory stress.

Oxidative stress was induced by exposing the cells to hydrogen peroxide (H_2O_2, 500 μM for 24 h), while inflammation was stimulated using lipopolysaccharide (LPS, 1 μg/mL for 24 h).

Control groups included untreated cells, stress-induced cells without curcumin, and curcumin-only groups to ensure that the observed effects were specifically related to the insult–treatment interaction.

Cell viability was measured using the MTT assay. Intracellular ROS generation was quantified using the DCFH-DA fluorescent probe, and mitochondrial membrane potential was assessed using the JC-1 dye.

Lipid peroxidation was determined by malondialdehyde quantification, while antioxidant enzyme activities, including superoxide dismutase, catalase, and glutathione peroxidase, were measured using spectrophotometric kits. Pro-inflammatory cytokine expression levels (TNF-α, IL-6, and IL-1β) were analyzed in culture supernatants by ELISA, and the expression of signaling proteins involved in inflammation and oxidative stress (e.g., NF-κB, MAPKs, and Nrf2) was evaluated by Western blotting.

4.2 *Ex vivo*-isolated skeletal muscle model

The *ex vivo* approach allows a more physiologically relevant assessment of herbal bioactives on skeletal muscle [64].

In this protocol, the extensor digitorum longus (EDL) or soleus muscle is dissected from anesthetized rodents and maintained in a Krebs–Henseleit buffer (95% O_2/5% CO_2, 37 °C).

Muscles are incubated with bioactive compounds (e.g., polyphenols, flavonoids, alkaloids) for 30–60 minutes prior to oxidative insult induced by tert-butyl hydroperoxide (t-BHP, 100–200 μM) or LPS (1 μg/mL).

Oxidative stress is evaluated by measuring lipid peroxidation (MDA levels), protein carbonylation, and glutathione redox status, while inflammatory responses are quantified by cytokine release and myeloperoxidase activity.

Mitochondrial function can also be assessed by measuring ATP content and mitochondrial membrane potential.

This model provides insights into the direct protective effects of herbal bioactives on intact muscle tissue.

4.3 *In vivo* studies

Animal models of muscle injury or strenuous exercise reveal that supplementation with turmeric, ginseng, or green tea extracts decreases inflammatory markers, reduces oxidative stress, and improves muscle performance.

Bioactives also accelerate muscle regeneration by modulating satellite cell activity and apoptosis pathways.

In vivo studies remain the gold standard for evaluating the pharmacological relevance of plant bioactives.

A commonly used protocol involves male Wistar rats or C57BL/6 mice subjected to muscle injury and inflammation through exercise-induced stress (treadmill running, eccentric contraction) or intraperitoneal injection of LPS (5 mg/kg).

Animals are divided into control and treatment groups, receiving bioactive substances orally (50–200 mg/kg/day) or intraperitoneally for 2–4 weeks.

Blood and muscle samples (gastrocnemius, soleus) were collected for biochemical analyses.

Oxidative stress biomarkers include MDA, 8-OHdG, and antioxidant enzyme activity (SOD, CAT, GPx).

Inflammatory responses are assessed by ELISA for TNF-α, IL-6, IL-1β, and by Western blotting for NF-κB and MAPK signaling.

Histological evaluation using H&E staining and immunohistochemistry for inflammatory markers (CD68, iNOS) provide morphological evidence of tissue protection.

This approach offers a comprehensive evaluation of the systemic and local effects of herbal bioactives on muscle health.

5. Example of therapeutic applications of bioactive compounds for skeletal muscle protection

5.1 Sports and exercise medicine

Athletes are often exposed to repeated bouts of exercise-induced muscle damage. Herbal bioactives, such as curcumin, ginger, and resveratrol, help reduce inflammation, accelerate recovery, and enhance performance.

5.2 Aging and sarcopenia

Sarcopenia, characterized by progressive muscle loss, is exacerbated by chronic inflammation and oxidative stress. Ursolic acid, resveratrol, and apigenin show promise in counteracting these processes by promoting mitochondrial function and reducing catabolic signaling.

5.3 Metabolic disorders

In obesity and type-2 diabetes, skeletal muscle is a key site of insulin resistance. Bioactives with anti-inflammatory and antioxidant effects improve insulin sensitivity and glucose uptake.

5.4 Rehabilitation and muscle injury

During recovery from injury, herbal bioactives support tissue regeneration, reduce fibrosis, and enhance healing outcomes.

6. Therapeutic modes of applications

Several modes of application have been explored for their therapeutic relevance:

- Oral administration (dietary or supplement form): Oral delivery is the most common mode, ensuring systemic bioavailability. Polyphenols such as curcumin, resveratrol, and quercetin, as well as omega-3 fatty acids, have been administered orally to modulate oxidative stress, mitochondrial biogenesis, and inflammatory signaling in muscle tissues. This approach is particularly used for the long-term management of sarcopenia, cachexia, and exercise-induced muscle damage [97].

- Parenteral administration (intraperitoneal, intravenous, and subcutaneous): Parenteral routes are employed mainly in experimental models to achieve higher plasma concentrations and bypass first-pass metabolism. Curcumin, epigallocatechin gallate (EGCG), and certain alkaloids have demonstrated improved therapeutic efficacy when delivered intraperitoneally in preclinical studies of muscle atrophy and dystrophy [63, 98].

- Topical and localized delivery: Bioactive-loaded hydrogels, patches, or creams can be applied directly to muscle or joint regions, providing localized action. This is particularly relevant in sports medicine to reduce oxidative and inflammatory damage following exercise or injury [14, 99].

- Nanoparticle and liposomal delivery systems: Advanced strategies, such as nanoparticle and liposomal formulations, improve solubility, stability, and targeted delivery of poorly bioavailable compounds, including curcumin and resveratrol. Nano-formulations enhance tissue penetration, prolong circulation time, and increase accumulation in skeletal muscle, showing promise for muscular dystrophies and degenerative conditions [100].

- Dietary fortification and functional foods: Incorporation of bioactive-rich extracts (e.g., flavonoids, carotenoids, plant sterols) into functional foods or beverages serves as a preventive approach to maintain muscle health and delay age-related sarcopenia [101].

- Gene and epigenetic modulation by nutraceuticals: Certain compounds, such as sulforaphane and resveratrol, act as epigenetic modulators, influencing transcriptional programs involved in muscle regeneration, mitochondrial function, and autophagy. This strategy provides long-term preventive and restorative potential [68].

- Combination therapies: Co-administration of bioactive compounds with exercise regimens, standard pharmacological agents, or physical rehabilitation strategies enhances therapeutic outcomes [102].

7. Conclusion

Herbal bioactive compounds represent a promising therapeutic avenue for managing skeletal muscle disorders associated with oxidative stress and inflammation. Numerous studies have demonstrated that polyphenols, flavonoids, alkaloids, and terpenoids exert protective effects on muscle tissues by neutralizing reactive oxygen species, modulating inflammatory pathways such as NF-κB and NLRP3, and improving mitochondrial function. These effects contribute to enhanced muscle recovery, reduced damage following strenuous exercise, and potential benefits in chronic conditions such as sarcopenia, cachexia, and muscular dystrophies.

Acknowledgments

The author confirms the use of AI-based tools for linguistic editing of the manuscript.

Conflict of Interest

The authors declare no conflict of interest.

Author details

Ahmed Karim[1*], Saliha Mharchi[2], Mohamed Marghich[3] and Abdelhay Addous[1]

1 Laboratory of Bioresources, Biotechnology, Ethnopharmacology, and Health, Faculty of Sciences, Mohammed First University, Oujda, Morocco

2 Water, Environment and Health team, LAPABE Laboratory, Faculty of Science, Mohammed First University, Oujda, Morocco

3 Nutritional Physiopathology, Neurosciences, and Toxicology Team, Laboratory of Anthropogenetics, Biotechnology, and Health, Faculty of Sciences, Chouaib Doukkali University, El Jadida, Morocco

*Address all correspondence to: a.karim@ump.ac.ma

References

[1] Kwee BJ, Mooney DJ. Biomaterials for skeletal muscle tissue engineering. Current Opinion in Biotechnology. 2017;**47**:16–22. DOI: 10.1016/j.copbio.2017.05.003

[2] Baskin KK, Winders BR, Olson EN. Muscle as a "mediator" of systemic metabolism. Cell Metabolism. 2015;**21**(2):237–248

[3] Gomarasca M, Banfi G, Lombardi G. Chapter Four-Myokines: The endocrine coupling of skeletal muscle and bone. In: Makowski GSBTA in CC, editor. Elsevier; 2020. p. 155–218. Available from: https://www.sciencedirect.com/science/article/pii/S0065242319300678

[4] Hoffmann C, Weigert C. Skeletal muscle as an endocrine organ: The role of myokines in exercise adaptations. Cold Spring Harbor Perspectives in Medicine. 2017;**7**(11):a029793

[5] Zhang H, Qi G, Wang K, Yang J, Shen Y, Yang X, et al. Oxidative stress: Roles in skeletal muscle atrophy. Biochemical Pharmacology. 2023;**214**(March):115664. DOI: 10.1016/j.bcp.2023.115664

[6] Tu H, Li YL. Inflammation balance in skeletal muscle damage and repair. Frontiers in Immunology. 2023;**14**:1–14

[7] Tidball JG. Inflammatory processes in muscle injury and repair. American Journal of Physiology-Regulatory, Integrative and Comparative Physiology. 2005;**288**(2):345–353

[8] Anbazhagan AN, Priyamvada S, Alrefai WA, Dudeja PK. Pathophysiology of IBD associated diarrhea. Tissue Barriers. 2018;**6**(2):1–21. DOI: 10.1080/21688370.2018.1463897

[9] Ananta MF, Saha P, Spriha SE, Rahman SMA, Chowdhury AKA. Analgesic, Anti-inflammatory and antidiarrheal properties of synthesized benzimidazole derivatives via in Vivo and in silico methods. Dhaka University Journal of Pharmaceutical Sciences. 2024

[10] Al-Bawardy B, Codipilly DC, Rubio-Tapia A, Bruining DH, Hansel SL, Murray JA. Celiac disease: A clinical review. Abdominal Radiology. 2017;**42**(2):351–360

[11] Tidball JG. Regulation of muscle growth and regeneration by the immune system. Nature Reviews Immunology. 2017;**17**(3):165–178

[12] Powers SK, Jackson MJ. Exercise-induced oxidative stress: Cellular mechanisms and impact on muscle force production. Physiological Reviews. 2008;**88**(4):1243–1276

[13] He F, Li J, Liu Z, Chuang CC, Yang W, Zuo L. Redox mechanism of reactive oxygen species in exercise. Frontiers in Physiology. 2016;**7**(NOV):1–10

[14] Karati D, Varghese R, Mahadik KR, Sharma R, Kumar D. Plant Bioactives in the Treatment of Inflammation of Skeletal Muscles: A Molecular Perspective. Evidence-based Complementary and Alternative Medicine. 2022;**20**:1–18 DOI: 10.1155/2022/4295802

[15] Mthembu SXH, Dludla PV, Ziqubu K, Nyambuya TM, Kappo AP,

Madoroba E, et al. The potential role of polyphenols in modulating mitochondrial bioenergetics within the skeletal muscle: A systematic review of preclinical models. Molecules. 2021;**26**:1–14

[16] Vargas-Mendoza N, Madrigal-Santillán E, Álvarez-González I, Madrigal-Bujaidar E, Anguiano-Robledo L, Aguilar-Faisal JL, et al. Phytochemicals in skeletal muscle health: Effects of curcumin (from Curcuma longa Linn) and Sulforaphane (from Brassicaceae) on muscle function, recovery and therapy of muscle atrophy. Plants. 2022;**11**:1–24

[17] Ahmed M. Targeting aging pathways with natural compounds: A review of curcumin, epigallocatechin gallate, thymoquinone, and resveratrol. Immunosenescence and Aging. 2025;**22**(1):28. DOI: 10.1186/s12979-025-00522-y

[18] Jang WY, Hwang JY, Cho JY. Ginsenosides from panax ginseng as key modulators of NF-κB signaling are powerful anti-inflammatory and anticancer agents. International Journal of Molecular Sciences. 2023;**24**(7):6119. DOI: 10.3390/ijms24076119

[19] Kramer HF, Goodyear LJ. Exercise, MAPK, and NF-κB signaling in skeletal muscle. Journal of Applied Physiology. 2007;**103**(1):388–395

[20] Liu T, Zhang L, Joo D, Sun SC.NF-κB signaling in inflammation. Signal Transduction and Targeted Therapy. 2017;**2**(e17023):1–9

[21] Mahdy MAA. Glycerol-induced injury as a new model of muscle regeneration. Cell and Tissue Research. 2018;**374**(2):233–241

[22] De Paepe B.A recipe for myositis: Nuclear factor κB and nuclear factor of activated T-cells transcription factor pathways spiced up by cytokines. AIMS Allergy and Immunology. 2017;**1**(1): 31–42

[23] Rayavarapu S, Coley W, Kinder TB, Nagaraju K. Idiopathic inflammatory myopathies: Pathogenic mechanisms of muscle weakness. Skeletal Muscle. 2013;**3**(1):13. DOI: 10.1186/2044-5040-3-13

[24] Huang L, Li M, Deng C, Qiu J, Wang K, Chang M, et al. Potential therapeutic strategies for skeletal muscle atrophy. Antioxidants. 2023;**12**(1):44. DOI: 10.3390/antiox12010044

[25] Jackson MJ, Pollock N, Staunton CA, Stretton C, Vasilaki A, McArdle A. Chapter 34-Oxidative Stress in Skeletal Muscle: Unraveling the Potential Beneficial and Deleterious Roles of Reactive Oxygen Species GERMAN. In: Sies HBTOS, editor. Academic Press; 2020. p. 713–733. Available from: https://www.sciencedirect.com/science/article/pii/B9780128186060000341

[26] Jackson MJ. Free radicals generated by contracting muscle: By-products of metabolism or key regulators of muscle function? Free Radical Biology and Medicine. 2008;**44**(2):132–141

[27] Sakellariou GK, Lightfoot AP, Earl KE, Stofanko M, McDonagh B. Redox homeostasis and age-related deficits in neuromuscular integrity and function. Journal of Cachexia, Sarcopenia and Muscle. 2017;**8**(6):881–906

[28] Powers SK, Duarte J, Kavazis AN, Talbert EE. Reactive oxygen species are signalling molecules for skeletal muscle adaptation. Experimental Physiology. 2010;**95**(1):1–9

[29] Ferreira LF, Reid MB. Muscle-derived ROS and thiol regulation in muscle fatigue. Journal of Applied Physiology. 2008;**104** (3):853–860

[30] Li LJ. Antioxidants and oxidative stress in exercise (44453). Experimental Biology and Medicine. 2000;**222** (44453):283–292.

[31] Fulle S, Protasi F, Di Tano G, Pietrangelo T, Beltramin A, Boncompagni S, et al. The contribution of reactive oxygen species to sarcopenia and muscle ageing. Experimental Gerontology 2004;**39**:17–24

[32] Kim JG, Sharma AR, Lee YH, Chatterjee S, Choi YJ, Rajvansh R, et al. Therapeutic potential of quercetin as an antioxidant for bone-muscle-tendon regeneration and aging. Aging and Disease 2025;**16**:1414–1437

[33] Le Moal E, Pialoux V, Juban G, Groussard C, Zouhal H, Chazaud B, et al. Redox control of skeletal muscle regeneration. Antioxidants and Redox Signaling 2017;**27**:276–310

[34] Dogan SA, Giacchin G, Zito E, Viscomi C. Redox signaling and stress in inherited myopathies. Antioxidants and Redox Signaling. 2022;**37** (4–6):301–323. DOI: 10.1089/ars.2021.0266

[35] Morgan MJ, Liu ZG. Crosstalk of reactive oxygen species and NF-κB signaling. Cell Research. 2011;**21** (1):103–115

[36] Nathan C, Cunningham-Bussel A. Beyond oxidative stress: An immunologist's guide to reactive oxygen species. Nature Reviews Immunology. 2013;**13**(5):349–361. DOI: 10.1038/nri3423

[37] Petersen MC, Shulman GI. Mechanisms of insulin action and insulin resistance. Physiological Reviews. 2018;**98**(4):2133–2223

[38] Li M, Qin Y, Geng R, Fang J, Kang SG, Huang K, et al. Effects and mechanisms of phytochemicals on skeletal muscle atrophy in glucolipid metabolic disorders: Current evidence and future perspectives. Food Innovation and Advances 2025;**4**:83–98

[39] Heinrich M. Fundamentals of Pharmacognosy and Phytotherapy. 2nd ed. Elsevier, editor. Vol. **17**. Churchill (London): British Library Cataloguing; 2012, p. 3–9

[40] Calixto JB. The role of natural products in modern drug discovery. Anais Da Academia Brasileira de Ciencias. 2019;**91**:1–7. DOI: 10.1590/0001-3765201920190105

[41] Liguori I, Russo G, Curcio F, Bulli G, Aran L, Della-Morte D, et al. Oxidative stress, aging, and diseases. Clinical Interventions in Aging. 2018;**13**:757–772

[42] Zeiger E, Møller IMMA. Plant Physiology and Development Oxford. 6th ed. United Kingdom (or New York, NY): Press OU; 2015, p. 1–888

[43] Wink M. Modes of action of herbal medicines and plant secondary metabolites. Medicines. 2015;**2** (3):251–286

[44] Scalbert A, Johnson IT, Saltmarsh M. Polyphenols: Antioxidants and beyond. American Journal of Clinical Nutrition. 2005;**81**(1 Suppl):215–217

[45] Li Y, Liu Y, Tan R, Liu Y. Effect of flavonoids on skeletal muscle mass, strength and physical performance in

middle-aged and older adults with or without Sarcopenia: A meta-analysis of randomized controlled trials. Frontiers in Nutrition. 2022;**9**(1):1013449

[46] Li AN, Li S, Zhang YJ, Xu XR, Chen YM, Li HB. Resources and biological activities of natural polyphenols. Nutrients. 2014;**6** (12):6020–6047

[47] Tsao R. Chemistry and biochemistry of dietary polyphenols. Nutrients. 2010;**2**(12):1231–1246

[48] da Silva W, Machado ÁS, Souza MA, Mello-Carpes PB, Carpes FP. Effect of green tea extract supplementation on exercise-induced delayed onset muscle soreness and muscular damage. Physiology and Behavior. 2018;**194**:77–82. DOI: 10.1016 /j.physbeh.2018.05.006

[49] Goh KP, Lee HY, Lau DP, Supaat W, Chan YH, Koh AFY. Effects of resveratrol in patients with type 2 diabetes mellitus on skeletal muscle SIRT1 expression and energy expenditure. International Journal of Sport Nutrition and Exercise Metabolism. 2014;**24**(1):2–13

[50] Kaur R, Arora S. Alkaloids-important therapeutic secondary metabolites of plant origin. Journal of Critical Reviews. 2015;**2**(3):1–8

[51] Wu Y, Yang Y, Du C, Peng X, Fan W, Chang B, et al. Berberine attenuates obesity-induced skeletal muscle atrophy via regulation of FUNDC1 in skeletal muscle of mice. Scientific Reports 2025;**15**:1–12

[52] Vega RJS, Xolalpa NC, Castro AJA, González CP, Ramos JP, Gutiérrez SP. Terpenes from Natural Products with Potential Anti-Inflammatory Activity. In: Perveen S, Al-Taweel A, editors. Terpenes and Terpenoids [Internet], Rijeka: IntechOpen; 2018. p. 60–85. DOI: 10.5772/intechopen.73215

[53] Kunkel SD, Elmore CJ, Bongers KS, Ebert SM, Fox DK, Dyle MC, et al. Ursolic acid increases skeletal muscle and brown fat and decreases diet-induced obesity, glucose intolerance and fatty liver disease. PLoS One. 2012;**7**(6):e39332

[54] Gupta SC, Patchva S, Aggarwal BB. Therapeutic roles of curcumin: Lessons learned from clinical trials. AAPS Journal. 2013;**15**(1):195–218

[55] Aggarwal BB, Sung B. Pharmacological basis for the role of curcumin in chronic diseases: An age-old spice with modern targets. Trends in Pharmacological Sciences. 2009;**30**(2):85–94

[56] Li Y, Yao J, Han C, Yang J, Chaudhry MT, Wang S, et al. Quercetin, inflammation and immunity. Nutrients 2016;**8**:1–14

[57] Sekkouma K, Cheritia A, Talebb S, Bourmitaa Y, Belboukharia N. Traditional phytotherapy for urinary diseases in Bechar district (South west of Algeria). Electronic Journal of Environmental, Agricultural and Food Chemistry. 2011;**10**(8):2616–2622

[58] Pisoschi AM, Pop A. The role of antioxidants in the chemistry of oxidative stress: A review. European Journal of Medicinal Chemistry. 2015;**97**:55–74. DOI: 10.1016/j. ejmech.2015.04.040

[59] Giovannini C, Filesi C, D'Archivio M, Scazzocchio B, Santangelo C, Masella R.[Polyphenols and endogenous antioxidant defences:

Effects on glutathione and glutathione related enzymes]. Annali dell'Istituto Superiore Di Sanità. 2006;**42**(3): 336–347

[60] Farkhondeh T, Folgado SL, Pourbagher-Shahri AM, Ashrafizadeh M, Samarghandian S. The therapeutic effect of resveratrol: Focusing on the Nrf2 signaling pathway. Biomedicine and Pharmacotherapy. 2020;**127**:110234

[61] Sin TK, Yu AP, Yung BY, Yip SP, Chan LW, Wong CS, et al. Effects of long-term resveratrol-induced SIRT1 activation on insulin and apoptotic signalling in aged skeletal muscle. Acta Diabetologica 2015;**52**:1063–1075

[62] Lagouge M, Argmann C, Gerhart-Hines Z, Meziane H, Lerin C, Daussin F, et al. Resveratrol improves mitochondrial function and protects against metabolic disease by activating SIRT1 and PGC-1α. Cell 2006;**127**: 1109–1122

[63] Lee DY, Chun YS, Kim JK, Lee JO, Lee YJ, Ku SK, et al. Curcumin ameliorated oxidative stress and inflammation- related muscle disorders in c2c12 myoblast cells. Antioxidants 2021;**10**:1–16

[64] McArdle A, Van Der Meulen JH, Catapano M, Symons MCR, Faulkner JA, Jackson MJ. Free radical activity following contraction-induced injury to the extensor digitorum longus muscles of rats. Free Radical Biology and Medicine. 1999;**26**(9–10): 1085–1091

[65] Yu T, Dohl J, Wang L, Chen Y, Gasier HG, Deuster PA. Curcumin ameliorates heat-induced injury through NADPH oxidase – dependent redox signaling and mitochondrial preservation in C2C12 myoblasts and mouse skeletal muscle. Journal of Nutrition. 2020;**150**(9):2257–2267. DOI: 10.1093/jn/nxaa201

[66] Lee DY, Chun YS, Kim JK, Lee JO, Lee YJ, Ku SK. Curcumin ameliorated oxidative stress and inflammation-related muscle disorders in C2C12 myoblast cells. Antioxidants. 2021;**10**(476):1–15

[67] Shabani M, Sade A, Hosseini H, Teimouri M. Resveratrol alleviates obesity- induced skeletal muscle inflammation via decreasing M1 macrophage polarization and increasing the regulatory T cell population. Scientific reports. 2020;**10**:1–13

[68] Qin X, Niu W, Zhao K, Luo Y, Wang W, He Y, et al. Resveratrol enhances post-injury muscle regeneration by regulating antioxidant and mitochondrial biogenesis. Current Research in Food Science. 2025;**10** (100972):1–10. DOI: 10.1016/j.crfs.2025.100972

[69] Sadeghi A, Seyyed Ebrahimi SS, Golestani A, Meshkani R. Resveratrol ameliorates palmitate-induced inflammation in skeletal muscle cells by attenuating oxidative stress and JNK/NF-κB pathway in a SIRT1-independent mechanism. Journal of Cellular Biochemistry. 2017;**118** (9):2654–2663

[70] Luk HY, Appell C, Chyu MC, Chen CH, Wang CY, Yang RS, et al. Impacts of green tea on joint and skeletal muscle health : prospects of translational nutrition. Antioxidants. 2020;**9**:1–28

[71] Meador BM, Mirza KA, Tian M, Skelding MB, Reaves LA, Edens NK. The

green tea polyphenol epigallocatechin-3-gallate (EGCG) attenuates skeletal muscle atrophy in a rat model of sarcopenia. Journal of Frailty and Aging. 2015;**4**(4):209–215

[72] Mandal S, Mitra A, Bose B, Shenoy PS. Quercetin and kaempferol mitigate endotoxin-induced skeletal muscle wasting by inhibiting KLF15 expression and restoring the antioxidant system. FASEB Journal. 2025;**39**(e70790). DOI: 10.1096/fj.202501184R

[73] Otsuka Y, Miyamoto N, Nagai A, Izumo T, Nakai M, Fukuda M, et al. Effects of quercetin glycoside supplementation combined with low-intensity resistance training on muscle quantity and stiffness. Frontiers in Nutrition. 2022;**9**(July):912217

[74] Chun C, Jai-Sing Y, Lu CC, Chiu YJ, Chen HC, Chung MI, et al. Effect of quercetin on dexamethasone-induced C2C12 skeletal muscle cell injury. Molecular Articles 2020;**25**:1–16

[75] Akiyama M, Mizokami T, Miyamoto S, Ikeda Y. Kaempferol increases intracellular ATP content in C2C12 myotubes under hypoxic conditions by suppressing the HIF-1 α stabilization and/or by enhancing the mitochondrial complex IV activity. Journal of Nutritional Biochemistry. 2022;**103**:108949. DOI: 10.1016/j.jnutbio.2022.108949

[76] Wang D, Yang Y, Zou X, Zhang J, Zheng Z, Wang Z. Antioxidant apigenin relieves age-related muscle atrophy by inhibiting oxidative stress and hyperactive mitophagy and apoptosis in skeletal muscle of mice. Journals of Gerontology: Biological Sciences. 2020;**75**(11):2081–2088

[77] Choi WH, Son HJ, Jang YJ, Ahn J, Jung CH, Ha TY. Apigenin ameliorates the obesity-induced skeletal muscle atrophy by attenuating mitochondrial dysfunction in the muscle of obese mice. Molecular Nutrition & Food Research. 2017;**61**(12):1700218. DOI: 10.1002/mnfr.201700218

[78] Chieko S, Tomoki A, Nobuhiko K, Ayako O. Flavones inhibit LPS-Induced Atrogin-1/MAFbx Expression in Mouse C2C12 skeletal myotubes. Journal of Nutritional Science and Vitaminology. 2015;**61**:188–194

[79] Böttcher N, Suhr F, Pufe T, Wruck CJ, Fragoulis A. Luteolin induces Nrf2 activity in C2C12 cells : Implications for muscle health. Molecular Sciences. 2025;**26**(4092):1–17

[80] Wang T, Zhang L, Chen S, Wang Z. Naringin Ameliorates IL-6 Mediated mitochondrial dysfunction and skeletal muscle degeneration via AMPK/Nrf-2 pathway. Research Square. 2024;**29**:1–23

[81] Termkwancharoen C, Malakul W, Phetrungnapha A, Tunsophon S. Naringin ameliorates skeletal muscle atrophy and improves insulin resistance in high-fat-diet-induced insulin resistance. Nutrients. 2022;**14**:4120

[82] Lv Z, Meng J, Yao S, Lv Z, Meng J, Yao S, et al. Article Naringenin improves muscle endurance via activation of the Sp1-ERR g transcriptional axis ll Naringenin improves muscle endurance via activation of the Sp1-ERR g transcriptional axis. Cell Reports. 2023;**42**(11):113288. DOI: 10.1016/j.celrep.2023.113288

[83] Jia Y, Wu C, Kim YS, Yang SO, Kim Y, Kim JS. et al. A dietary

anthocyanin cyanidin-3-O-glucoside binds to PPARs to regulate glucose metabolism and insulin sensitivity in mice. Communications Biology. 2020;**3** (514):2–11. DOI: 10.1038/s42003-020-01231-6

[84] Scazzocchio B, Varì R, Filesi C, Archivio MD, Santangelo C, Giovannini C, et al. Cyanidin-3- O - b - glucoside and protocatechuic acid exert insulin-like effects by upregulating PPAR g activity in human omental adipocytes. Diabetes 2011;**60**: 2234–2244

[85] Geng X, Ji J, Liu Y, Li X, Chen Y, Su L. et al. Cyanidin-3-O-glucoside supplementation ameliorates metabolic insulin resistance via restoration of nitric oxide-mediated endothelial insulin transport. Mol Nutr \& Food Res. 2022;**66**(2):2100742. Available from: https://onlinelibrary.wiley.com/doi/abs/10.1002/mnfr.202100742

[86] Lv X, Zhao XP, Li WC, Xing NF, Zong KQ, Zhai Y, et al. Anthocyanins and musculoskeletal diseases: Mechanisms and therapeutic potential. Frontiers in Microbiology. 2025;**12**:1–28

[87] Kimble R, Jones K, Howatson G. The effect of dietary anthocyanins on biochemical, physiological, and subjective exercise recovery: A systematic review and meta- analysis. Food Science and Nutrition. 2023;**63** (9):1262–1276

[88] Li Y, Chen X, He J, Zheng P, Luo Y, Yu B, et al. Grape seed proanthocyanidin extract promotes skeletal muscle fiber type transformation through modulation of cecal microbiota and enhanced butyric acid production. Journal of Food Science. 2024;**89**(6):3788–3801. DOI: 10.1111/1750-3841.17075

[89] Hah YS, Lee WK, Lee SJ, Lee SY, Seo JH, Kim EJ, et al. Rutin prevents dexamethasone-induced muscle loss in C2C12 myotube and mouse model by controlling. Antioxidants. 2023;**12**:1–11

[90] Poudel A, Zhou JY, Mekala N, Welchko R, Rosca MG, Li L. Berberine hydrochloride protects against cytokine-induced inflammation through multiple pathways in undifferentiated. Canadian Journal of Physiology and Pharmacology. 2019;**97** (8):699–707

[91] Lin KH, Chu WX, Lee YT, Li YH, Chang WT, Yu YP, et al. Effects of sweet potato leaf extracts and chlorogenic acid on glucose uptake in C2C12 Cells. Agronomy. 2024;**14**:1–12

[92] He J, He Z, Wang H, Zhang C, Pei T, Yan S. et al. Biomedicine & pharmacotherapy caffeic acid alleviates skeletal muscle atrophy in 5/6 nephrectomy rats through the TLR4/MYD88/NF-kB pathway. Biomedicine and Pharmacotherapy. 2024;**174** (116556):1–15. DOI: 10.1016/j.biopha.2024.116556

[93] Jeong MJ, Lim DS, Kim SO, Park C, Leem SH, Lee H. et al. Protection of oxidative stress-induced DNA damage and apoptosis by rosmarinic acid in murine myoblast C2C12 cells. Biotechnology and Bioprocess Engineering. 2022;**27**(2):171–182. DOI: 10.1007/s12257-021-0248-1

[94] Moraes FDSR, Moraes LHR, Macedo AB, Valduga AH, Mizobuti DS, Rocha GLD. et al. Anti-inflammatory and antioxidant activities of Citrus aurantifolia (Christm.) Swingle essential oil in dystrophic muscle cells: Implication of the PGC-1α pathway. Journal of Essential Oil-Bearing Plants.

2024;**27**(1):34–46. DOI: 10.1080/0972060X.2023.2296994

[95] Tomoya K, Naoki H, Hidetaka I, Ryoichi Y. b -carotene increases muscle mass and hypertrophy in the soleus muscle in mice. Journal of Nutritional Science and Vitaminology. 2015;**61**:481–487

[96] Liu S, Yang D, Yu L, Aluo Z, Zhang Z, Qi Y et al. Effects of lycopene on skeletal muscle-fiber type and high-fat diet-induced oxidative stress. Journal of Nutritional Biochemistry. 2021;**87**:108523. DOI: 10.1016/j.jnutbio.2020.108523

[97] Capozzi A, Saucier C, Bisbal C, Lambert K. Grape polyphenols in the treatment of human skeletal muscle damage due to inflammation and oxidative stress during obesity and aging: Early outcomes and promises. Molecules. 2022;**27**(19):6594

[98] Receno CN, Liang C, Korol DL, Atalay M, Heffernan KS, Brutsaert TD, et al. Effects of prolonged dietary curcumin exposure on skeletal muscle biochemical and functional responses of aged male rats. International Journal of Molecular Sciences. 2019;**20**(5):1178

[99] Yoon W, Kim J. Effect of topical sports cream containing cymbopogon schoenanthus on acute recovery markers following exercise-induced muscle fatigue in competitive athletes: a pilot study. Journal of Men's Health. 2022;**18**(10):198

[100] Yadav A, Yadav SS, Singh S, Dabur R. Natural products: Potential therapeutic agents to prevent skeletal muscle atrophy. European Journal of Pharmacology. 2022;**925**(September 2021):174995. DOI: 10.1016/j.ejphar.2022.174995

[101] Li Q, Yang H, Song S, Liu J, Wang Z, Wang J. Bioactive components in whole grains for the regulation of skeletal muscle function. Foods. 2022;**11**(18)

[102] Nyakayo AO. Exploring medicinal plant synergies: combination therapies for disease management. Newport International Journal of Research in Medical Sciences. 2025;**6**(2):120–128